QoS
Measurement and Evaluation of Telecommunications Quality of Service

QoS Measurement and Evaluation of Telecommunications Quality of Service

William C. Hardy
WorldCom, USA

JOHN WILEY & SONS, LTD
Chichester · New York · Weinheim · Brisbane · Singapore · Toronto

Other Wiley Editorial Offices

John Wiley & Sons, Inc., 605 Third Avenue,
New York, NY 10158-0012, USA

WILEY-VCH Verlag GmbH
Pappelallee 3, D-69469 Weinheim, Germany

John Wiley & Sons Australia, Ltd, 33 Park Road, Milton,
Queensland 4064, Australia

John Wiley & Sons (Canada) Ltd, 22 Worcester Road
Rexdale, Ontario, M9W 1Ll, Canada

John Wiley & Sons (Asia) Pte Ltd, 2 Clementi Loop #02-01,
Jin Xing Distripark, Singapore 129809

British Library Cataloguing in Publication Data

A catalogue record for this book is available from the British Library

ISBN 0471499579

Typeset in Times by Deerpark Publishing Services Ltd, Shannon, Ireland
Printed and bound in Great Britain by Biddles Ltd, Guildford and King's Lynn, UK

This book is printed on acid-free paper responsibly manufactured from sustainable forestry, in which at
least two trees are planted for each one used for paper production.

For Adriana

Contents

Preface

Most people know that quality of service (QoS) in telecommunications has grown in importance over the past decade. This is thanks to the new competitive environment which has followed as a direct result of privatization and de-regulation, forcing companies to increase the quality of their networks and services. Yet QoS means different things to different people. In some developing countries where it is a struggle for QoS managers to wrestle with outdated equipment, even making a network perform in the way it was designed is an improvement in QoS.

The Quality of Service Development Group (QSDG) is a field trial group of QoS professionals from over 130 carriers, service providers, research companies and vendors from around the world. While informal, we operate under the auspices of Study Group 2 of the ITU-T. We gather annually in different geographic regions to discuss QoS issues within our companies. *QSDG Magazine* (www.qsdg.com) which as well as being our group's official magazine, is also the only periodical in the world about QoS, and is distributed in 201 countries and territories.

William C. "Chris" Hardy is unquestionably among the leading lights in the field of QoS. As chairman of the QSDG I appreciate the contributions Chris has made, both to the QSDG group as a whole, and through his *QDSG Magazine* column *Telecom Tips and Quality Quandaries*, on which much of this book is based. If you are coming to grips with QoS in your company, this is the place to start.

Luis Sousa Cardoso
QSDG Chairman
VU/Marconi
Lisbon, Portugal
January, 2001

Foreword

My involvement in analysis of quality of telecommunications services began almost by accident in June, 1967, when I started my first full-time job out of graduate school. The job was with the Operations Evaluation Group of the Center for Naval Analyses. It seems that what they happened to need the day I reported was someone to fill a slot as a communications analyst. Since I was there, I was anointed, never mind that I knew absolutely nothing about tele-communications systems, electrical engineering, or even electricity, since I had skipped that part of the college physics curriculum, and almost nothing of my graduate education in mathematics was relevant to understanding Navy tactical voice and teletype communications over radio frequency channels.

Because my career started with such a complete lack of practical experience and technical skills, my analytical efforts have never been marred or impeded by technical expertise or conventional wisdom. Rather, what I discovered was that all I really needed to do to be effective as a problem solver in this area was to:

- Imagine myself using the system I was studying;
- Decide what I would be concerned about if I were using it;
- Research the technology of the system to the extent necessary to understand the mechanisms affecting performance of the system with respect to those concerns; and
- Formalize the relationships between system performance and user perception of quality of service gleaned from this drill.

When I did this, everything else needed to solve the problem would readily follow – the user view would suggest concerns; concerns would suggest measures of quality and effectiveness; understanding of the mechanisms would suggest measures of performance and their relationship to measures of quality; measures would suggest quantifiers; quantifiers would suggest data requirements; and so on, all the way down the analytical chain.

This book is based on more than 30 years experience in successfully applying this approach in analyzing issues of quality of service of telecommunications systems to produce practicable solutions to quality problems. Because of the very basic nature of the approach, this book is apt to be viewed by some as being short on technical content and long on formulation of evaluative concepts and generic measures. However, I refuse to apologize for this, because the perspectives on quality of telecommunications services that I am trying to lay out here are exactly those that I would want all of my employees to share, were I ever to become the CEO of a telecommunications company, so that, for example:

- My marketing and sales forces would know how to communicate with customers in a way that would demonstrate their understanding of customers' concerns;
- My system engineers would know how to design my networks to satisfy customer expectations, rather than simply meet industry design standards;
- My operations managers would know the comfortable levels of performance affecting quality of services that must be achieved and maintained to assure user satisfaction;
- My service technicians would know how to troubleshoot user complaints with the same competence that they identify, diagnose, and correct technical problems; and
- Everyone involved anywhere in the company would have a very good idea of exactly how their day-to-day activities affect user perception of the quality of our services.

To this end, what I have tried to present here is a treatise on the ways and means of measuring and evaluating telecommunications services that is simple and straightforward enough to be appreciated by anyone, but sophisticated enough to be informative and useful to telecommunications professionals. The only way you can judge whether I have succeeded is to turn the page…

William C. Hardy
WorldCom, USA

Introduction

The purpose of this book is to define and describe a family of measures of quality of telecommunications services that have been demonstrated in their successful application over many years to be useful both to telecommunications service users, as a basis for understanding and assessing possible differences between competing services, and to service providers, as a means of determining what improvements in service performance are needed to assure customer satisfaction. The distinguishing characteristic of these measures is that they have in every instance been designed to simultaneously achieve two ends:

1. The credible, reliable assessment of the likelihood that users will find a particular service to be satisfactory; and
2. The determination of how system performance must be changed when that assessment shows that users are not likely to be satisfied.

This kind of complementary utility in a measurement scheme is not hard to achieve. However, it is, in fact, frequently absent in proposed quality of service (QoS) metrics, because definition and development of particular measures have failed to take into account both the concerns of the users of telecommunications services and the perspectives of the engineers and technicians who must design, build, and operate the systems that deliver those services. It is, therefore, a secondary, but equally important objective of this book to describe the analytical perspectives and discipline that have reliably guided the development of the specific measures that are presented here.

To this end, the material in this book is divided into two parts:

- Part I presents the concepts and perspectives that have guided the development of the measures. This section first presents what might be thought of as a theory of measurement. It begins with an examination of the possible reasons for developing measures and proceeds with a formal description of the process by which the measures discussed here were developed. This part of the book also contains a chapter that briefly defines and describes basic telecommunications functions and the processes by which those functions are used to deliver telecommunications services.

- Part II then discusses a complete family of measures of QoS of telecommunications services, keyed to the user concerns and different types of telecommunications services defined in Part I.

Under this organization of the material, then, Part II comprises the source material that can be researched for specific measures and applications, while Part I comprises both the background necessary to follow the development of the particular measures, and the "how to" manual for those who may be called upon to develop measures of QoS for new services or new ways of delivering services.

This structure allows for a variety of approaches to the material.

Persons who are conversant with telecommunications services and QoS measurement may choose to begin with Part II, and then revert to Part I for purposes of understanding the perspectives that supported development of the measures. Alternatively, a seasoned QoS analyst might read through Part I and readily acquire an understanding of the analytical discipline and techniques sufficient for purposes of developing measures for new services that are useful both to service users and to telecommunications system operators and engineers. Finally, persons with lesser background and experience in QoS will find that reading Part I first to get the grounding in the basics will make it much easier to follow the reasoning that justifies the selection of the measures described in Part II as being particularly well-suited for purposes of measuring and analyzing the particular aspect of QoS each describes.

Whatever the background and experience of the reader, I hope that this book shall clearly convey, both by force of reasoning and by example, three principles to be applied in defining and developing measures of QoS:

1. *Meaningful* measurement of quality of a telecommunications service must begin with a consideration of the concerns of the users of that service to develop a set of evaluative concepts that will guide the definition of measures and measurement schemes,
2. *Useful* measurement of QoS must be based on measures that can be readily interpreted by users, but are also clearly related to the performance characteristics of the systems that deliver the service, and
3. *Cost-effective* measurement of QoS can be realized only when the means of quantifying or estimating any measure is consciously selected on the basis of consideration of both the intended use of the measure and readily available sources of data.

Part I

Basic Concepts

1

Definitions

The subject of this book is quality of telecommunications services. Its focus is defining measures of quality of service (QoS) that can be used to evaluate telecommunications services in ways that are operationally meaningful, useful to decision-makers, and which can be achieved with a minimum investment in time and money.

Any readers who are comfortable with the description above can go directly to Part II. However, for those for whom this description, like *Jabberwocky* to Alice, evokes the reaction: "Somehow it seems to fill my head with ideas – only I don't exactly know what they are!" I shall initiate this journey by playing Humpty-Dumpty and explaining some of the more overworked words.

1.1 Quality of Service

In the present case the 'service' in the term 'quality of telecommunications service' is understood to pertain to something that is provided day-to-day for the use of someone, referred to throughout this book as a *user* of that service. As such, a telecommunications service is a particular capability to communicate with other parties by transmitting and receiving information in a way that is fully specified with respect to: how the user initiates a transaction; the mode in which the information is exchanged; how the information is formatted for transmission; how end-to-end exchanges of the information are effected; and how the transactions completed are billed and paid for. The important distinction in concept between the service and the systems or capabilities that deliver it is that users, as opposed to providers of the service, experience and care about only those characteristics of the service that are manifested when they try to effect the end-to-end communications transactions.

The 'quality' in 'quality of telecommunications service' is a much more elusive concept, for which neither any of the Websters nor Lewis Carroll can provide much help. The closest dictionary definition is "excellence of character", and if there are two meanings packed into one of Humpty-Dumpty's "portmanteau words" the term 'quality' in modern parlance carries a whole train load of loosely coupled meanings that are wont to head off in their own directions at any time.

The problem is that 'quality' as it is commonly understood in the context of 'quality of service' is "something" by which a user of the service will judge how good the service is. And, that something is expressed in the singular, making it synonymous with 'excellence' or 'grade', depending on whether it is viewed as what ought to be or actually is, respectively. In truth, however, 'quality' in this context is very plural. The factors that will determine how highly a user rates QoS are inescapably multidimensional, both with respect to the attributes of the service that the user will value, and the perspectives on the service, which will determine what is appropriately graded to gauge likely user assessment of value.

To appreciate the multidimensional nature of the attributes of service that users will value, imagine yourself trying to sell a telephone service that is otherwise excellent in all respects, but is horribly deficient in some aspect. Your sales spiels might run something like this:

- We guarantee that our service will always be there and ready to go when you want to use it. So we just do not see how you can possibly be worried about that little 1-min call set-up time problem.
- 99.95% of calls placed with our service will result in a connection! And, only 50.7% of those connections will be to the wrong number.
- 99.9% of calls placed with our service will result in the right connection! Now, we understand that there might be some difficulty in hearing each other, when the connection is up, but...

The point is that there are many possible attributes of service that may shape a user's perception of quality. These attributes are, moreover, independent, so that inability to meet user expectations with respect to any one of them cannot be offset by exceeding user expectations with respect to the others, any more than stylish design of an automobile chassis and a nicely appointed leather interior can off-set a poor engine design that makes the car a gas guzzling maintenance nightmare. In practical terms, this means that effective measurement of QoS will necessarily involve a collection of measures, rather than "the" measure of QoS, to serve as a basis for gauging likely user perception of service quality.

The other complication of the notion of 'quality' is one of perspective. The essential distinctions are illustrated in the simplified model shown in Figure 1.1, which comes out of a briefing from about 1982. Some of the descriptions

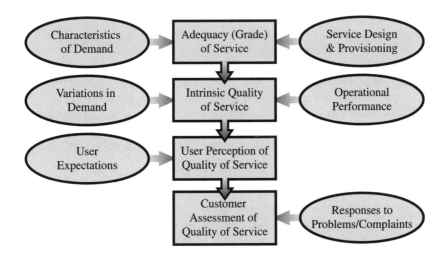

Figure 1.1 Simplified model of factors that shape perception of quality of service

in the boxes have been changed to conform to modern terminology, and a lot has been left off, but the thrust of the message remains the same. When you look at the factors that will determine whether a customer will buy a particular telecommunications service and stay with it, there are at least three distinct, but interrelated notions of "quality of service" that might come into play in the evaluation:

- The first is what might be thought of as an *intrinsic* quality of service. Such intrinsic quality is achieved via:

  - The technical design of the transport network and terminations, which determine the characteristics of the connections made through the network, and
  - Provisioning of network accesses, terminations, and switch-to-switch links, which determines whether the network will have adequate capacity to handle the anticipated demand.

Since the goal is to be able to implement within that network various telecommunications services whose quality should be competitive in the target marketplace, intrinsic service quality is usually gauged by expected values of measures of operational performance characteristics and verified by demonstration that those scores compare favorably with analogous scores of competing services.

- The second notion of quality of a particular service is what might be called *perceived* quality of service. Perceived quality results when the service is actually used, at which time the users experience the effects of intrinsic service quality on their communications activities, in their environment, in handling their demand, and react to that experience in light of their personal expectations. As suggested in Figure 1.1, those expectations are usually conditioned by users' experience with similar telecommunications services, but may also be influenced by representations by the service vendor as to how the service will compare to others with which a user may be familiar.
- The third level of quality can be thought of as *assessed* quality of a particular service, which results when the user/customer who pays for the service makes the determination whether the quality of service is good enough to warrant its continued use. As shown in Figure 1.1, this notion of quality of service depends directly on the perceived quality of service, but is also affected by other considerations, principal among which are the vendor responses to problems with the service.

The importance of these distinctions seen as follows.

1.1.1   Intrinsic vs. Perceived Quality of Service

The distinction of the notions of perceived and intrinsic quality of service is a critical one, because it is perceived, rather than intrinsic, quality that ultimately determines whether a user will be satisfied with the service delivered. This was the painful lesson that we learned when I worked at Satellite Business Systems, back in the 1980s. By all common measures of clarity of voice services, the satellite links offered much higher intrinsic voice quality. There was less signal attenuation, less noise, and no higher incidence of perceptible echo over the satellite circuits than was occurring over comparable terrestrial routes. However, there were differences in characteristics that were not commonly measured, such as the crystal clarity of echo, super quiet connections that made people think that a call had been disconnected when the distant party stopped talking, and longer transmission delays, that were causing some users to experience discomfiture with the satellite service when it replaced the terrestrial service with which they were familiar. As a consequence, perceived quality of service was in this case at variance with the indications from analysis of intrinsic quality of service, demonstrating that measures of intrinsic quality of service alone can be useless as a basis for predicting user satisfaction.

Or, consider the deceptively simple question of adequacy of post-dial delay

(PDD). The intrinsic quality of a particular service with respect to PDD is pretty much set by the design of the underlying network, depending, for example, on how calls are routed; whether dialed digits are translated for switching; how variable length numbers are handled; and extent to which node-to-node signaling to set up connections is effected via in-band digit spill, rather than out-of-band, digital link signaling. Consideration of the particulars for any type of route will therefore pretty much define what PDD will be achieved, and a "safe" basis for determining whether a particular service will be competitive will be a demonstration that the PDD experienced over any type of route will not be appreciably different from the least PDD over that type of route achieved by competing vendors.

Beyond this, however, users/customers who are sophisticated enough to recognize that there may be a very wide range of PDDs among different vendors' offerings of a particular service will demand some representation from competing vendors as to "average" PDDs or other information that will address the direct concern: "If I buy your service, will I/my user community experience unacceptably long post-dial delay?" Because this question addresses the issue of perceived, rather than intrinsic, quality with respect to PDD, there are two pitfalls in relying only on the values describing intrinsic PDD.

The first is that the measures of intrinsic PDD can be accurate only to the extent that the different types of routes actually used by the target community replicate the distribution of different types of routes over the network. Thus, for example, without considering the particulars of usage of the proposed service, a vendor can easily wind up telling someone whose international calls are all destined for rural areas of outer Mongolia to expect a large percentage of those calls to have the 2 s PDD achieved in trans-oceanic calls between countries with modern all-digital domestic networks.

Worse yet is the fact that even a very accurate description of measures of intrinsic PDD to be expected by a user will still be useless in predicting user satisfaction unless there is some medium for reliably determining what will be an unacceptable long PDD for the target user community. Without such a translator of the measurements used to gauge intrinsic PDD, vendors may feel compelled to develop and offer the least PDDs afforded by current technology, possibly leading to a situation in which the vendor community has gone to great lengths to be able to offer delays that are, say, less than a second to a user community that really does not care about PDDs as long as they are no greater than the 6 s to which they have become accustomed, and may even be bothered by unexpectedly fast network responses.

1.1.2   Perceived vs. Assessed Quality of Service

Viewed in another way, intrinsic quality of service is what may make a particular service attractive to a buyer in the first place, but perceived quality of service is what will determine whether that buyer will find the service acceptable when it is delivered. In contrast, what we refer to here as 'assessed' quality of service is what will determine whether the buyer will retain the service or dump it at the first opportunity. The first requirement for good assessed quality of service is, of course, that the perceived quality of service is acceptable to the user community. However, there are other factors that can result in an unsatisfactory assessment of a particular a service whose perceived quality of service is completely acceptable...

...such as when that service produces a spontaneous disconnect of a phone call between the president of a company and a very important client just as the president is about to clinch a deal, inducing the president to demand immediate change of the service, regardless of cost (true story, though cooler heads prevailed over the "regardless of cost" condition).

...such as the otherwise acceptable service that is dropped, because a customer service representative treated the user like an imbecile and became abusive and insulting when the user persisted in trying to explain the problem (everyman's story).

...such as the otherwise acceptable service for which the bill for one line for 1 month was erroneously posted as $1000...and the vendor's accounts representative refused to correct it...and the vendor turned the overdue bill over to "Your Money or Your Knees" collection agency (everyone's nightmare).

In terms of the preceding discussions of the meaning of QoS, the most important measures of interest will be those that enable us to describe in quantitative terms perceived quality of service in ways that will relate directly to intrinsic quality of service, and to identify in qualitative terms those service characteristics that will affect the determination of assessed quality, with respect to essential sets of service attributes that will shape user perception of quality. The descriptions of these measures will in each case represent the application of analytical perspectives that have been successfully applied over the last 30-odd years to facilitate selection and definition of measures. Because the measures described in Part II cannot be easily rationalized or described without appeal to the resultant models, the following sections focus on those perspectives, beginning with a particular view of what measurement of anything is all about, and concluding with definitions of generic telecommunications functions and the systems that will be repeatedly used in describing measures of QoS in Part II.

2

Measurement and Evaluation

At the beginning of this section, it was declared that the focus of this book will be definition of *measures* of QoS that can be used to *evaluate* telecommunications services in ways that are operationally meaningful, useful to decision-makers, and achieved with a minimum investment in time and money. As used in this book, the italicized terms refer to the end products of what are conceived as two distinct processes. The first, *measurement*, is one which produces quantitative descriptions of attributes of a telecommunications service that affect the user perception of its quality; the second, *evaluation*, is one whereby those quantitative descriptions are interpreted to answer some specific question, such as whether users can be expected to be satisfied with a particular service, what might be done to improve user satisfaction, or whether users might find some change in intrinsic quality to be worthwhile.

Taken together, these two processes comprise what might commonly be thought of as an *analysis* of QoS. The reason for explicitly recognizing and distinguishing the two processes involved is that far too often the measurement of QoS is thought of as the end of the analysis, rather than a necessary step en route to producing the evaluations that provide specific answers to specific questions. Such a perception of analysis of QoS fosters a number of altogether pernicious notions, such as: the idea that needs for analysis of QoS can be met by generating routine reports of measurements; the view that it is the job of the QoS analyst to dream up some complicated expression for producing a single measure of "quality" that reflects everything or typifies "quality" across all regions where a particular service is provided; and the common misconception that some measures of intrinsic quality of service are adequate surrogates for measures of perceived quality of service.

2.1   Function of Measurement and Evaluation

What is suggested here, then, is that analysis is a process whose ultimate end is to produce specific answers to specific questions. This point of view is predicated on the modest assertion that:

> **The only good reason to measure anything is to reduce uncertainty with respect to some course of action that must be decided.**

Admittedly, this statement has some of the flavor of the Caterpillar trying to tell Alice which is the right and left side of a round mushroom. However, all that is posited here is that measurement and evaluation to produce and interpret quantitative descriptions of performance, quality, or whatever other attributes are being examined, will neither be useful nor worthwhile unless the results help someone feel more comfortable about some decision as to what to do and when to do it, such as what new car to buy, what telephone services to order, how to go about correcting a recognized problem, how to recognize that a problem has emerged, or when to sell a stock. Without such an underlying need for the information gleaned from measurement and evaluation, the results will be of no more use to a decision-maker than a painstaking analysis of carefully collected data showing with great precision and confidence that the sun will nova in exactly 9 787 316 years, 3 months, and 4.7 h, evoking responses from decision-makers that the results are "interesting", or more damning, "nice-to-know", but not "actionable".

The principal value of this concept of the function of measurement and evaluation is that it readily suggests a number of questions that the analyst should address before undertaking any analysis. These include questions of:

- *Audience*: which decision-makers are to be supported by the results of the analysis?
- *Utility*: what kinds of decisions are to be facilitated? How must measurements be evaluated to produce information that can be used for those decisions?
- *Concerns*: what are the questions that those decision-makers are likely to want to have answered during the course of making those decisions?
- *Objectives*: what are the courses of action that will be decided or determined by appeal to the results of the analysis?

2.1.1   Audience and Utility

To appreciate the importance of addressing these questions at the outset, consider first the diversity of possible audiences for analyses of quality of telecommunications services. As described below, there are at least five distinct classes of decision-makers who might be responsible for actions whose efficacy depends on reliable information of likely user perception of QoS, and the evaluation of measures needed to make the results of the analysis useful to the decision-makers is in each case different.

(1) *Service users.* The most obvious class comprises the actual users of the service, who are continually testing its quality by placing calls. The principal uncertainties that they face are ones of how often they will encounter problems that materially impede the act of placing a call and completing the desired exchanges of information. Consequently, users will be very conscious of any difficulties experienced and will synthesize that experience over time to determine whether the incidence and severity of problems actually encountered is acceptable, thereby producing a subjective assessment of perceived quality. On the basis of that subjective assessment, a user then decides tentatively that the service is satisfactory or unsatisfactory. If it is unsatisfactory, the user will initially complain, and then later abandon the service, if the is no improvement. If the service is tentatively found to be satisfactory, the user continues its use and continues to synthesize the experience with it to verify the original subjective assessment. As long as the assessment does not change, the user remains satisfied. However, perceptible changes in the type, incidence, severity, or user's accommodation of problems with the service may result in a different assessment of perceived quality, leading the user to decide to complain about or change the service, when possible. As a possible audience for results of QoS analyses, then, users will be looking for results providing reassurances with respect to uncertainties as to what will be experienced in the unknown future. Such reassurances sought will be of one of two kinds:

- Assurances that a service that has not been experienced, such as a new offering, a less expensive substitute for an existing service of the same kind, or a similar service based on new technology is likely to be found to be satisfactory; or
- Assurances that a service that has been experienced and found to be unsatisfactory will be put right and no longer exhibit the type, severity, or incidence of problems that rendered it unsatisfactory in the first place.

Since users are the ultimate decision-makers with respect to which of possibly many competing services is to be used, the user concerns are the principal focus of QoS measurement, and the evaluation of those measures should answer the basic question:

What is the likelihood that users of a service exhibiting the value x for the QoS measure M_p, will find the service to be satisfactory with respect to the attribute measured by M_p?

(2) *User representatives.* Users of residential and small business telecommunications services usually represent themselves in such activities as selecting telecommunications services and features, choosing among competing providers of the chosen services, and negotiating prices. However, such activities are otherwise vested in a small group of people whose principal decision-maker, whom we will call the Comm Manager, is responsible for choosing, acquiring, and maintaining services for a large body of users. Since Comm Managers are the representatives of their user communities, they must be concerned with user satisfaction with the services they select, and are therefore naturally interested in analyses of perceived quality of service as a means of reassuring their users of the validity of their decisions. However, since their role is also one of assuring their management of economy of services, their perspective on QoS will be one of trying to assess cost-benefit trade-offs, and the principal question with respect to measures of QoS will frequently be more like:

> What is the smallest value x for the QoS measure M_p that will keep complaints from my user community as to the quality of service with respect to the attribute measured by M_p at manageable levels?

In addition, by virtue of being the principal decision-maker for a user community, the Comm Manager will be the one responsible for the assessed quality of service. The Comm Manager will therefore be much more concerned with questions of billing and customer support, and much more actively involved in trying to define and assure satisfaction of the criteria for assessed quality, than the individual user.

(3) *Service provider sales and marketing personnel.* On the other side of the fence, one of the major consumers of QoS analyses will be the sales and marketing personnel, who are not necessarily decision-makers, but must respond to the concerns with QoS raised by the users and Comm Managers who are their prospective customers. Because of their role in telling prospective customers about telecommunications services, they will want whatever the customer wants, but with the additional feature that the analyses must also show how quality of the services they sell compares with that of competing services offered by other providers. Because of the need to characterize, communicate, and interpret any differences in measures of QoS between the competing telecommunications services, their principal questions with respect to evaluation of QoS is usually (or by all means should be):

> What does the difference between the value x for the QoS measure M_p for the service we sell and the value y for a competing service really mean to users? Will it be noticeable? Will any noticeable differences be great enough to alter the users' synthesis of their experience to produce an assessment of perceived QoS?

(4) *Service operations and maintenance personnel.* Standing right behind the sales and marketing personnel, usually cursing them for creating unrealistic customer expectations of QoS, are the service provider's operations and maintenance personnel, who are responsible for monitoring day-to-day performance of the systems that deliver the service to assure that QoS is maintained at acceptable levels. Because they must be able to understand and act on QoS via actions taken on those aspects of operations that are within their control, their focus is necessarily on intrinsic quality of service, and their principal questions with respect to measurement and evaluation of QoS will be ones of the relationship between measures of intrinsic and perceived QoS of the form:

What values of the measure of intrinsic QoS, M_i, will indicate likely user dissatisfaction with the perceived quality of the attributes of service of concern to users affected by the characteristic of operational system performance measured by M_i?

Analysis of perceived QoS, then, will be largely worthless to operations and maintenance decision-makers unless the evaluation of the results is extended to produce derived *indicators* of specific conditions that must be corrected in order to avoid deleterious effects on the service users' assessment of perceived quality.

(5) *System architects and engineers.* Last on our list of possible consumers of QoS measurement and evaluation are the persons who must make the decisions as to the technology to be employed in implementing various telecommunications services and the way various assets are to be configured to deliver particular services. Like operations and maintenance personnel, the system architects and engineers are concerned with intrinsic quality. Unlike operations and maintenance personnel, who are constrained to manage performance within the constraints of the existing system and resources, the architects and engineers are responsible for deciding the characteristics of the telecommunications system and the allocation of resources that will achieve intrinsic quality adequate to assure a high likelihood that perceived quality will be acceptable. To do this, they must have hard and fast requirements that can be used as the basis of system design and configuration. Notions of subjectivity and perception must be totally factored out of the equations, and the fuzzy indicators that might be used for operations and maintenance management must be replaced by *criteria* for acceptability of variations of intrinsic quality that are technical, concrete, specific and completely unambiguous. The need for such criteria, then, generates questions of the form:

What value, x, of the measure of intrinsic QoS, M_i, is an upper/lower limit for what must be achieved in the system design to assure the ability to deliver acceptable perceived QoS?

2.2   More Definitions

Consideration of questions of audience for, and utility of, a particular analysis thus begins to shape our perceptions of what kinds of measurements should be taken and how they are to be evaluated in order to best serve the needs of the intended audience. Examination of concerns similarly helps shape our perception of what characteristics of the service should be described and quantified for the analysis, while consideration of objectives will suggest the most efficient means of quantifying those characteristics from available data.

Before describing how this happens, however, it is necessary to take time out to pay some words and hire some others to mean "just what I choose them to mean, neither more nor less". The workforce so far is shown in Table 2.1. Some of these words have already been defined implicitly, so their definitions should by now be reasonable and understandable, but probably would have caused your eyes to glaze over had I laid them out in that way for you earlier. The newcomers are: data, information, measures, quantifiers, concerns and objectives. They are defined in Table 2.1 so that we can make the following distinctions.

2.2.1   Data vs. Information

One of the most pernicious practices in the world of telecommunications is that of treating the problem of analysis of QoS as one of gathering up some of the readily available data that abounds in our data-rich environment, throwing it into a database management system to provide capabilities for database query in order to enable users to "drill down" or do "data mining", adding some statistical summarization algorithms and graphing capabilities to detect and display "trends", and reducing the question of purpose of such analysis to one of deciding what reports and displays to produce. Such a malconception of the nature of analysis creates the baseless expectation that decision-makers' questions can be answered by generating reports from such systems, without the added dimension of evaluation.

To make it clear that such systems cannot be expected to suffice as a means of analysis of QoS, the definitions of "data" and "information" set forth in Table 2.1 draw a clear distinction between the products from analysis by asserting, in essence, that the necessary product of measurement is *data*, while the desired product of evaluation is *information*. Information thus becomes something extracted from data that answers specific questions so as to reduce uncertainty. Anything else, no matter how elegantly summarized, or beautifully displayed in charts and graphs, is still just data.

The resultant distinction between what can be called "data" and what will

Table 2.1 Definitions of some of the unconventionally defined words used in this book

Telecommunications service	A set of capabilities provided to a user that enables the user to set up and effect exchanges of information to a distant destination
Quality of service (QoS)	An answer to the question: "How well does a particular service perform relative to expectations?" The type of quality involved may be distinguished as being:
– Intrinsic	Relative to the expectations of the persons who design and operate the systems that deliver the telecommunications service;
– Perceived	Relative to the expectations of the persons who use the service; or
– Assessed	Including the expectations of the persons who must deal with the providers of the service on matters of billing, ordering, correction of problems, etc.
QoS thus depends on *whose* expectations are the basis for gauging quality	
Measurement	A process by which a numerical value is assigned to some attribute of an entity under examination
Concern (with service)	An uncertainty as to whether what will be experienced with respect to some attribute of a service will meet expectations
Measure	A description of some attribute of an entity that can be expressed as a number or quantity; used everywhere in this book to refer to *what* is described
Quantifier	A definition of the variables and calculations that are to be used to compute the value of a measure
Evaluation	A process by which values of measures are interpreted to reduce uncertainties; evaluations of quality reduce uncertainties as to whether what will be experienced with respect to some attribute of service will meet expectations
Objective	The purpose of an evaluation, as described by the nature of the decision(s) that will be supported
Indicator	A quantifier of a measure that is useful when the objective of the evaluation is to determine whether a particular event or condition has occurred
Criterion (pl. criteria)	A basis for evaluation of a measure expressed as a single value (threshold) which is used to assign an acceptable/unacceptable rating depending on whether the value of the measure is above or below the threshold

Table 2.1 (*continued*)

Data	A collection of facts, observations, or measurements that might be used in assigning a value to a measure
Information	The results of interpretation of data to produce answers to specific questions whose answers will effectively reduce uncertainty with respect to a decision that must be made

be called "information" becomes a valuable criterion for the quality of an analysis of QoS. Unless the results from that analysis can be fairly labeled "information" in the sense of the definition in Table 2.1, you can be pretty sure those results will not satisfy the intended audience.

2.2.2 Measures vs. Quantifiers

From the point of view adopted here, the desired product of the measurement process is a data set comprising numbers representing a set of measurements. To completely and accurately describe such a data set, it is necessary to detail precisely *what* has been measured and *how* the measurement has been made. As suggested by the definitions in Table 2.1, the distinction between the what and the how in a description of a data set is usefully captured by distinguishing between:

- *Measures*, which define what is to be described in quantitative terms without any restriction on what is to be calculated from data; and
- *Quantifiers*, which describe how the associated measure is to be (was) expressed as a quantity.

In this scheme of things, a measure then becomes the precise, unchanging definition of what should be expressed as a numerical value, while the actual numerical values in a data set may have been produced by reference to any number of quantifiers for that measure. The measure thus becomes the name of a quantification of a particular attribute of an entity being analyzed, such as "height" of a person, and the quantifier becomes an expression that specifies one of possibly many ways that a numerical value is to be assigned to that attribute, e.g. "height" as defined by the distance between the sole of the foot at the heel and the top of the head as measured in feet and inches between parallel planes containing these points.

To see the importance of making such distinctions, consider this example. As a measure of service quality in the sense defined in Table 2.1, "availability" can be understood to refer to an unspecified quantity that accurately describes expectations that the service will be fully functional and available for use when it is needed. An associated quantifier of that measure may also be called "availability", but "availability" in this case will refer to the ratio, MTBF/(MTBF + MTTR), or other, equivalent metrics derived from data, such as the ratio (total time the service was up and ready for use)/(total time the service was observed), or from estimates of the incidence and duration of outages, which can be used to calculate those ratios.

Although this distinction is made here for purposes of facilitating descriptions of tools and techniques for measurement of QoS, there are concrete benefits of such a seemingly esoteric, theoretical distinction. For example, this distinction removes any possibility of wasting time on those philosophical arguments as to the "proper" definition of a particular measure. If the measure has been well-defined, everyone can readily apprehend what we are talking about, and the question of which one of possibly many quantifiers of that measure to use can be decided by selecting the quantifier that makes the most cost-effective use of the data that can be readily acquired, without confusing its meaning or limiting our capability for the desired evaluation. Similarly, the distinction between measures and quantifiers naturally leads us to require a description of both the measure and the quantifiers for a set of measurements, thereby avoiding the common pitfall of trying to synthesize and evaluate measurements without consideration of how those measurements were made.

2.2.3   Concerns

If measures and quantifiers describe the what and how of measurement, then *concerns* explain the *why*. As suggested earlier, and made explicit in the definition in Table 2.1, the term "concern" is used here as the rubric for an uncertainty that must be addressed in the evaluation of measurements. To make them concrete, such concerns will usually be described as a set of questions posed as to the likelihood of occurrence of undesirable events or conditions.

In accordance with the perspective of purpose of measurements articulated earlier, it is the existence of those uncertainties that is the sole reason for conducting measurements. Consequently, there is such a natural, ready association between concerns and measures as defined in Table 2.1 that *the description of the concern nearly always defines the attribute to be measured*. Since what we refer to as measures are usually identified by naming the attribute to be measured, this means that there is usually little ambiguity in using the same name for the measure and the concern, thereby making this association explicit.

For example, consider the concern, expressed as a question: "Will the system be fully operational and available for use when I want to use it?" The reason that "availability" is identified as one of the important characteristics of a system is that the word "availability" is a good, intuitive one-word name for the system attribute that is the object of the concern expressed. And, it is as readily understood that a measure called "availability" would be something that could be used to answer that question in meaningful quantitative terms, expressing in this case the probability that the system will be available for use.

This suggests, and my experience proves, that a preliminary characterization of likely concerns of the intended audience for an analysis will lead almost unerringly to selection and definition of measures for that analysis that are readily understandable by, and meaningful to, the audience.

2.2.4   Objectives

Finally, if concerns as defined in Table 2.1 explain the reason for conducting an analysis, the *objective(s)* as defined there characterizes its *envisioned utility* to the intended audience. Note that in the sense of the word as it is used here the term "objective" does not refer to what the analyst is to accomplish, or what the analysis of QoS is to show. Rather, what is referred to as an objective of an analysis here is a description of the decisions to be made that generated the concerns to be addressed in the first place. Such objectives will, then, be properly described by completing the sentence: *The results of this analysis will be used in deciding/determining whether...by...*

The reason for insisting that the objectives of an analysis be couched in these terms is that it drives home the axiom put forth earlier that the only good reason to measure anything is to reduce uncertainty with respect to some course of action that must be decided. However, this particular definition of objectives also has a very practical benefit for analysis of QoS in that it complements the benefit from consideration of concerns. Just as a formal description of concerns serves as an automatic guide to selection of measures that will ensure an analysis of quality of service that is effective for its intended purpose, selecting quantifiers of those measures based on a clear understanding of the *objectives* of the analysis in light of the readily available sources of data will lead unerringly to the selection of the most cost-effective quantifiers for the defined measures.

To see what I mean by this, suppose someone requests an analysis of how long it takes for a call to complete through a particular service. Without consideration of the objectives of that analysis, the analyst is very likely to select as the quantifier for the analysis of the PDD, as measured by the following difference:

(time the first ring back signal or voice answer is detected)

   − (time the last digit was dialed)

This is a very precise quantifier of how long it takes to complete a call, but it is also sometimes very hard to acquire the data required to use it. Unless the service can be readily instrumented for automatic timing of call progress, the manual dialing and timing required to acquire adequate data may be daunting. And, even when it can be instrumented, there will still be the time- and labor-consuming activities such as shipping and installing test devices in appropriate locations, checking them out, writing and testing data collection scripts, etc. all of which is required to collect the data.

Now suppose that we add to that effort the question of the *objective* of measuring the PDD. Then some of the possible answers and their influence on the selection of the quantifier and consequent cost of obtaining the data might be the following:

- *The analysis will be used to determine when there has been significant change in the time it takes to complete a call.* In this case, there is probably more than enough data to satisfy the objective in the billing records for the service, which will show the time that circuits handling calls placed via the service were seized, together with the time that answer supervision was received for completed calls. These data will then support ready calculation of the *answer time* for completed calls, defined as the difference:

   (time of receipt of answer supervision)

      − (time the service access circuit was seized)

   This quantifier does not accurately estimate the PDD. However comparisons of the average answer times from the large, homogenous, stable samples from two different time periods that can be readily constructed from the billing records will reveal any significant changes in PDD.

- *The analysis will be used to decide whether the service is competitive with respect to time required to complete a call.* In this case, the evaluation can be based on measurements of the time to complete calls taken from services whose call handling is the same as the service in question, or by summing engineering estimates of time expected for the different steps in the call completion process. These estimates will be crude ones for the service in question, but they will be adequate quantifiers of the time required to complete a call for the objective of the analysis, because most users will be indifferent to differences that are much larger than the inaccuracies of the estimates. Notice also, however, that this objective mandates something that might have been overlooked – acquisition of commensurate measurements of the time it takes to complete a call for the competing services with which the service in question will be compared.

- *The analysis will be used to decide whether the call set-up process for the service in question is functioning properly and isolate any deterioration in performance.* In this case the only useful quantifier for the amount of time it takes to complete a call is an estimate resulting from the sum of observations of the time required at each step in the call completion process taken under different operating conditions. The overall PDD that might have been selected as the quantifier for the analysis and its variations might be useful in deciding when to look for service deterioration. However, this objective cannot be satisfied unless the engineering estimates of the time expected for different steps in the call completion process are supplanted by actual measurements of each step in the process that are far more fine-grained than can be achieved with the instruments that can sample PDD.

In each of these examples, then, there is the same, well-defined concern as to how long it takes to complete a call, which defines the associated measure. However, failure to consider the objective might in each case create the possibility of adopting for the analysis a quantifier for that measure than would either not support the objective or involve much more time and effort in acquiring the necessary data and quantifying the measure. This is why I have made a career of being an obnoxious obstructionist to measurement efforts by insisting that before deciding what data are to be accumulated, there are two questions that must be answered:

- Who is the likely audience (cognizant decision-maker)? and
- What are the objectives?

Needless to say, such insistence can sometimes make me very unpopular with those who would rather be getting down to the nitty-gritty of defining the databases that need to be created to measure QoS.

3

The Analysis Process

Like analysis itself, the process by which analyses comprising measurement and evaluation are conducted can be thought of as comprising multiple phases. The phases in this case are:

- *Formulation*, during which the audience, decisions supported, etc. are clarified and used as the basis for determining and specifying measurement requirements.
- *Data handling*, during which the data elements needed to quantify each measure are acquired, organized, and manipulated.
- *Evaluation*, during which values of the measures are calculated and interpreted as necessary to address the specific concerns of the intended audience.

3.1 Phase 1: Formulation

The earlier discussions of concepts of measurement and evaluation suggest a formal process that should be followed in structuring any analytical effort to assure that the end results will be operationally meaningful, useful to decision-makers, and achieved with a minimum investment in time and money. The principal steps in that process are described in Figure 3.1, which displays the relationships among the six principal steps of that process and the structure of an intermediate decision loop for selecting quantifiers.

The six steps are as follows.

3.1.1   Identify the Audience

As suggested in Figure 3.1, the recommended first step in formulating any analytical effort is to determine the intended users of its results. In Part II of this book, for example, the audience interested in the analysis of QoS is at the outset presumed to be the service users, whose proximate concerns are perceived QoS, and the development of measures and quantifiers are extended to serve the needs of other audiences whose principal concerns are with intrinsic or assessed QoS only where it appears to be useful. Whenever such extensions occur, it will be seen that the new measures discussed would seem

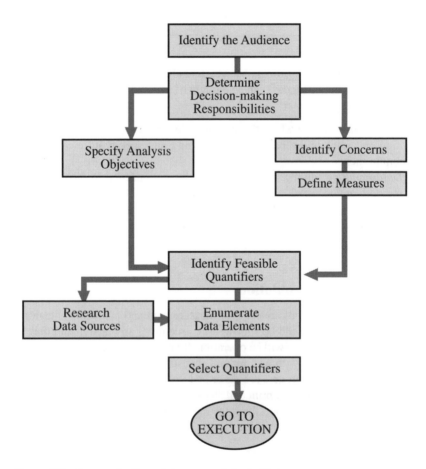

Figure 3.1   Process for formulating an analytical effort

to be wholly out of place without the explicit warning that there is a change in the intended audience.

3.1.2   Determine Decision-Making Responsibilities

Once the target audience is identified, the next step in the structured approach to formulating an analysis recommended here is a *conscious* determination of the decisions or general kinds of decisions that will be facilitated by its results. As suggested in earlier discussions of measurement and evaluation, those decisions will be some course of action with respect to the service, such as its purchase, continued use, marketing, operation and maintenance, or design.

For example, the basic user decision with respect to QoS is whether to keep the current service or shift to another. The alternative may simply be the same kind of service offered by a competing provider, or a new kind of service for meeting old requirements, such as wireless voice telephony, or a new technology designed to handle combinations of old requirements in new ways, such as ISDN or a wideband subscriber loop into the home to replace the analog loop. However, the basic decision to be made is always the same: *Should I stick with what I have or jump to something different?* Other kinds of decisions that may be facilitated by analysis of QoS for other audiences are suggested in Table 3.1.

3.1.3   Specify Analysis Objectives

As shown in Figure 3.1, a third step in the process, but not necessarily the third in order, is to review the decision-making responsibilities of the audience to formulate specific analysis objectives. For service users, for example, it was suggested earlier that an analysis of QoS should support the decision to buy or keep a particular service by producing results that will:

1. Enable users to determine that a service that has not been experienced will in all likelihood be found to be satisfactory; or
2. Reassure users that a service that has been experienced and found to be unsatisfactory will be put right and no longer exhibit the type, severity, or incidence of problems that rendered it unsatisfactory in the first place.

Other possible analysis objectives for other audiences are exemplified in Table 3.1.

3.1.4   Identify Concerns

Having identified the decision-makers comprising the target audience and the

Table 3.1 Typical uses for analysis of quality of service

Audience	Decision-making responsibility	Assist decision-maker in	By
Operations and maintenance	Selection of the best use of available resources	Setting performance goals	Demonstrating effect of different levels of performance on likely user perception of quality
Marketing	How best to respond to competition claims of superior performance	Formulating counter-claim strategy and advertising	• Showing how differences in performance affect perceived quality • Characterizing services in terms of user, rather than operational or engineering, concerns
Engineering	Design of new services and selection of new equipment	Setting design criteria and standards	Characterizing current intrinsic quality and demonstrating parameters that will reliably assure that future quality will be good enough to be competitive
Human resources	Training and orientation for customer service representatives	Deciding what to stress in describing user courses and lectures	Determining and describing user sensitivities that affect assessed vs. perceived quality
Strategic planners	Investments in expansion, enhancement and development of services	Evaluating new services and concepts of operations	Assessing their attractiveness to users and contribution to the company's competitive posture

decisions that will be supported, it is also necessary to consider that audience and articulate the specific uncertainties that are likely to impede decision-making. Those uncertainties have been defined here to be concerns, usually expressed in the form of questions that can be readily understood by almost anyone.

The importance of prefacing any definition of measures for the analysis with an enumeration of likely concerns cannot be over stressed, because it is the key to assuring that the measures will be meaningful to the intended audience and useful in decision-making. Consider, for example, the case of the service users, who will be presumed to be the principal audience for the measures of QoS developed in Part II. If we were to simply adopt the measures of QoS cited in analyses targeted for technically knowledgeable persons responsible for operational decision-making or evaluation of system technology, the results of analysis would not be likely to be convincing or helpful for the users' purposes of deciding what service to buy and how long to keep it. The reason is that users seldom buy, and frequently do not even understand, technology. Their perceptions of the quality of a telecommunications service are instead based on how well that service meets their expectations and satisfies their needs when they use it. Thus, if the users cannot readily tell from an analysis based on technical measures what to expect from day-to-day use of the service, the results of the analysis will simply replace one set of uncertainties to be resolved with other uncertainties that are even harder to resolve.

3.1.5   Define Measures

As suggested by the preceding observations and shown in Figure 3.1, then, the definition of measures to be used in any analysis effort should be deferred until the relevant concerns of the intended audience have been identified. This recommendation is often anathema to those who are looking for quick results. However, the time invested in the orderly formulation of the analysis will be amply rewarded by the ease with which useful, meaningful measures can be defined at this step. If the steps shown above this one in Figure 3.1 have been taken, the analyst should find that the generic measures needed for the analysis will be nearly automatically defined by simply defining the most general quantities that might be used in formulating answers to the concerns described. As indicated earlier, this effort should, moreover, be so intuitive and natural that the attributes of the service to be measured will probably be identified in the description of the concerns, and the name of that attribute can readily be applied both to the concern and the measure without ambiguity.

3.1.6   Select Quantifiers

Once the measures have been defined and the analysis objectives have been clarified, it is then a relatively easy matter to select the most cost-effective quantifier for each measure. The iterative steps ending in this part of the process are illustrated in Figure 3.1. and described as follows:

1. *Identify feasible quantifiers.* The objectives specified for the analysis will determine for each measure the possible acceptable forms or modes of quantifiers, determining, for example: whether a precise value is needed or an indicator will suffice; whether a commensurate value from another service is needed to support comparisons; whether it will be more efficient to quantify the measure directly, or by estimating it as a function of sub measures, to ensure the ability to relate observed values of the measure to measures of contributing factors that must be distinguished; etc.
2. Then, to assess the cost-effectiveness of each of the feasible quantifiers identified for each measure, each member of a set of acceptable quantifiers for a measure is considered in turn to:

   – *Enumerate data elements* required to assign a value to the measure in accordance with the definition of the quantifier; and
   – *Research data sources* to determine the ease with which the necessary data elements can be acquired.

3. On the basis of the assessments in step (2), the quantifier for each measure is selected as the one among the feasible quantifiers for which the data elements can be most easily or most quickly acquired, depending on whether speed or ease of production of the analysis is the greater concern in the context of its application for the intended audience.

3.1.7   Example

Appendix A contains an application of the analysis formulation process just described to the question: "How do we gauge the quality of a service whereby telecommunications capacity is provided as needed by a customer?" In this example the service is referred to as *on-call provisioning.* Because the essence of such a service is to provide ready back-up capacity for large capacity services, such as large private networks, the presumed audience comprises *user representatives,* who will be responsible for deciding whether to buy on-call provisioning services for a large set of users.

The presumed analysis objective is to produce information that can be used by user representatives to determine which, if any, of various different versions on-call provisioning will be of value to their user communities and to decide whether the expected benefits warrant the cost.

The principal user concern that might be addressed in evaluating on-call provisioning is identified as being one of responsiveness to problems of capacity shortfalls in their networks. Since, from the perspective of the user representative, the expected benefit to the users would be avoidance of major problems arising from capacity shortfalls, the measure of effectiveness of the service with respect to responsiveness was defined to be the expected proportion of the requests for additional capacity that will be met in time to avoid major problems. This makes the measure defined more meaningful in the context of deciding whether the service would be worth its cost rather than simply measuring the technical characteristic of how fast the service would respond. Also, since the analysis is envisioned to facilitate decision-making by different users, that basic definition is extended to a generic measure explicitly recognizing factors that might differ among different users and different versions of an on-call provisioning service.

Then, to facilitate the evaluation of costs of different versions or brands on on-call provisioning services, the formulation of the analysis provides a discussion of the cost factors that must be considered under different circumstances and advice to the decision-maker as to how to assess costs, but without doing any of the actual dollar accounting.

From the development of the description of the basis for the evaluation of the service, the formulation of the analysis then proceeds with selections of quantifiers for the measure of the responsiveness for a variety of different versions of the service. In each case the quantifier specified represents an estimate of the proportion of the time that service responsiveness would be fast enough to avert major problems by whatever criterion the buyer might gauge how long users would tolerate the condition(s) to be corrected. However, the formulas are different, because they have been chosen to utilize data elements that can reasonably be expected to be easily recognized and readily available in the context of the envisioned implementation and application of on-call provisioning.

The result is a virtual guidebook for determining what data should be collected and what quantifier of responsiveness should be used for measurement and evaluation of various different versions of on-call provisioning in various different environments. Note, also, in the structured formulation of the analysis shown in Appendix A that some of the quantifiers for responsiveness are fairly complex and might otherwise be rejected as too hard to understand by the audience, but now appear to be quite credible, because the detailing of the evaluative concepts resonates with the real-world perspectives of intended audience...or so I claim.

3.2 Phase 2: Data Handling

The formulation of the analysis produces a list of all of the data elements that are needed for the analysis, together with the selected sources of those data elements. It is then, and only then, that the analyst is ready to initiate the part of the analytical process that many analysts want to be its start – the fun stuff of acquiring, organizing, and manipulating of all of those numbers that will be used to produce values of measures in accordance with the formulas specified by the selected quantifiers.

The principal activities encountered on this side of the looking glass are:

- *Data acquisition.* Creation of data sets comprising the elements that are needed to quantify measures;
- *Data organization.* Sorting, tagging, and arranging the elements of the data sets to create coherent databases that can be readily queried for well-defined subsets of the data; and
- *Data manipulation.* To clean up data sets, quantify measures, and facilitate understanding of the variations in values of quantifiers under different conditions.

The following provides some perspectives on what these activities might entail and some helpful tips on what to do when you are up to your armpits in the morass.

3.2.1 Data Acquisition

The first job after formulating the analysis is to aggregate enough of the right kind of data to support meaningful analysis and interpretation of variations in values assigned to the measures selected for the analysis. If the analysis has been well formulated in accordance with the procedures just described, there will be no question of *what* data sets are to be created or from *where* the data are to be acquired, because each quantifier will be defined in terms appropriate for a particular data source that was selected in advance to be the best alternative from among the available sources. However, even when the data source is known, there is what seems to be a universal fixation on the question of *how much* data will be enough for purposes of the analysis.

There are several possible answers to this question, depending on how the data are acquired. For example, if the data source is an existing database that has been created by someone else and is regularly maintained, and that database has been examined and deemed to be the best source of the data elements that need to be sampled for purposes of the measurements required in the analysis, then the simple answer to the question of how much is "all of it". There is no penalty in querying the database to extract *every* relevant data element, or in writing programs to be executed by the host of the data to produce anything we need from the total database, so there is no need to worry about sample sizes.

At the opposite extreme are situations that occur, for example, when the data must be collected by means of a survey or some other labor intensive set of observations. In this case, there is a relatively high price attached to each data element, and it is meaningful to ask what is the least amount of data that will suffice for purposes of the analysis. The problem is the answer must be predicated on a precise definition of what it means for a sample size to "suffice".

3.2.2 All the Statistics You Need to Know to Read this Book

Suppose that we record N repeated samples of a quantity to create the set of individual values $\{X_i | i = 1,2,3...N\}$. Then the arithmetic average of these values, $\bar{X}$, calculated by setting:

$$\bar{X} = (1/N) \sum_{i=1}^{N} X_i$$

is called the *mean* of the values $\{X_i\}$. A common measure of the variation in the sampled values is the *standard deviation* of the sample, S, calculated by setting:

$$S = \sqrt{\left[(N)\left(\sum_{i=1}^{N} X_i^2 \right) - \left(\sum_{i=1}^{N} X_i \right)^2 \right] / [(N)(N-1)]}$$

which is just a convenient way of finding the square root of the average sum of the squares of $\{\bar{X} - X_i\}$, representing the set of differences between the sample values and the mean.

Now, if we want that sample size, N, to be "big enough", the "enough" has to be defined in terms of three values: the standard deviation expected for the sampling procedure (σ); the desired accuracy in the estimate ($\pm \delta$); and the confidence level for the estimate (α), representing the probability that $\bar{X} = v_a \pm \delta$, where v_a is the actual value of the quantity sampled.

When all this is known or decided, the desired sample size ($SS_{\min}$) can be then determined by setting:

$$SS_{\min} = \left[(n(\alpha))\left(\sigma^2 \right) \right] / \left(\delta^2 \right),$$

where $n(\alpha)$ is a multiplier determined by the value of α according to the following short table.

For α = confidence (%) of:	Set $n(\alpha)$ =
90	1.65
95	1.96
99	2.58
99.5	2.81

Moreover, when v_a is a proportion whose value is between 0 and 1, and we can make an educated guess that v_a is approximately equal to some value, P, then we can set:

$$SS_{min} = [(n(\alpha))(P)(1 - P)]/\left(\delta^2\right)$$

Application of statistical theory produces the equations like that shown above for estimating the required sample size. However, that equation and all others of its ilk are single equations with multiple unknowns, requiring us to provide in advance estimates of the other variables in the equation. Thus, for example, to use the equation above to estimate N = the smallest sample size that will suffice to assure adequate accuracy and confidence in the results, we need to specify four things:

- A rough estimate of v_a, the actual value of the measure we are trying to estimate
- The sample standard deviation (σ);
- The desired accuracy in the estimate ($\pm \delta$); and
- The confidence level for the estimate (α), representing the probability that the actual value will be within the desired bounds of accuracy.

Thus, any attempt to answer this question requires subjective assignments of values of α and δ, and a guess as to the values of v_a and σ before having seen the first data point. Appeal to the equation thus supplants the original uncertainty as to how much data to collect with other kinds of uncertainties that may be as, or more, difficult to resolve to the satisfaction of the audience for the analysis.

For this reason, the best answer to the question of how much data to collect when there is value in minimizing the number of data points that must be acquired is to "Wait and see what develops". Instead of trying to determine the necessary sample sizes in advance, it is far more fruitful to approach this problem strategically, by adopting, when possible, what I refer to as a *cascaded* sampling strategy.

Under such a strategy sampling proceeds in steps, with constant rechecking to determine the status of the sample size relative to whether the sample size is sufficient according to criteria like those just stated. First, some data is collected, and the results are summarized to replace the original guess of ν_a with any better guess supported by the data. Next, the original guess of σ is replaced by a new value estimated from the data. More data are then added to the sample and the refinement process is repeated, and so on.

As this process proceeds, one of two things will happen. Either the actual sample, or the actual sample size with a small addition will equal or exceed the calculated minimum sample size, or it will become apparent that the sample size needed to satisfy the originally specified goals with respect to accuracy and confidence will be prohibitively expensive. In the former case, any further samples necessary are added to the data set, and sampling ceases. If the latter condition is realized, then the values of ν_a and σ for the data already collected are used to determine how the minimum sample size is reduced by reduction of the confidence level or the desired accuracy of the estimate. This examination of possibilities will then result in a set of conditions that can be satisfied with the existing data set, or feasible additions thereto, that represent a reasonable compromise between cost and sufficiency of the data set, or a realization that it will take too much time and effort to use that kind of sampling to produce measurements that are tight enough to be useful. In the latter case, the "wait and see" approach has brought us to an impasse necessitating some reformulation of the analysis. However, we also have not collected 500 data points in good faith at $100 a pop, only to find that the best we can do is provide the decision maker with information that has a 50/50 chance of being wrong.

Intermediate to these extremes is the case of data collection where there is a moderate cost per data element, but there are a large number of different factors that might cause the variations in the data, so it looks as if an impossibly large sample size will be required to be able to determine the effects of the different factors. The classical model for this is the problem of conducting a national opinion poll on some question. Depending on the nature of the question, the state or area of the country where persons reside, sex, age and any number of other characteristics are seen to be likely influences on the answers to the question posed.

The way the pollster takes this into account, then, is to: (a) partition each factor into possibilities; (b) use those divisions to define categories of responders to the question; and (c) construct a sample for which the proportion of persons in each category in the sample is the same as the proportion of persons the population tested. Thus, for example, for the likely influences just named, the partitions might be each of the 50 states of the US, male or female, and age intervals of 10 years for persons aged 20–70. By this partitioning, there are $50 \times 2 \times 5 = 500$ categories, which would include, for example, New York/male/40–50 years old and California/female/20–30 years old. The pollster will then try to construct a sample in which the ratio of the number of persons polled in each category to the total sample size closely approximates the percentage of residents of the US between 20 and 70 years old who fall into each category.

Because the problem of the pollster is so familiar, this method of structuring a sample to reflect the population it came from is intuitively appealing. However, such a sampling strategy may be good when the only objective is to measure something, but it is absolutely *not* recommended for analyses comprising measurement and evaluation. The principal reason is that, at the end of the day, the pollster may be able to report the best practicable estimate of the percentage of people in the country who feel one way or another about a particular issue, but there may not be enough data different categories in that sample to answer the questions about differences, such as "do the people in Vermont have a substantially different opinion about the issue than the people in New Mexico?"

The preferred sampling strategy for analyses, then, is to create a sample that can be meaningfully disaggregated to answer questions about differences among constituents. This is accomplished by initially proceeding just as the pollster does, to identify the principal factors that might be expected to affect the attribute that we want to measure, partition each factor into well-defined possibilities, and use those partitions to define categories. However, once the categories are defined, the objective in creating a sample is to assure that:

1. There are the same number of randomly sampled observations in each category; and
2. The number of samples is large enough to assure that differences in the effects of each of the factors originally identified can be tested statistically and characterized by combining results from different categories.

Thus, for example, from a sample created according to these rules, the answer to the question about the comparison of opinions between people from Vermont and New Mexico can always be answered by comparing all of the results from Vermont with all those from New Mexico to determine whether there are possible differences in the sample as collected. Moreover, the pollsters' results that might be necessary to account for different demographics by sex and age can be readily calculated from the data in the sample by weighting the results in each category by either state by the appropriate proportion from the same state population statistics that the pollster used to create the sample that reflected those demographics.

If you have kept with me on this example, then, the reward is that there is a very unambiguous answer to the question of how much data is necessary for cases where there is a moderate cost of data acquisition. If you structure the sample in the way just suggested to define categories, the target sample size that I have found to be adequate for all of the limited sample size analyses that I have conducted is *15 samples per category*. [Note: statisticians will immediately recognize that the value of 15 here is exactly half the nominal minimum sample size of 30 used for the normal approximation of the binomial distribution; halving that number in this case merely represents a bet that we can find for any circumstance at least two categories that can be combined for purposes of testing the significance of the influence of a particular factor.]

When the sample size must be reduced, the criterion of at least 15 observations per category can be used to determine how many categories can be distinguished, leading the analyst to repartition the factors in a way that has the least impact on the ability to characterize the effects of the different factors on the data observed.

Another aspect to the question of how much data is needed that is particularly important arises when the sample is to be used for evaluation of QoS of a new service. In all cases in which small or moderate size samples are involved, the sampling plan should always include a provision for the collection of data from both the new service to be evaluated and the old, or a competing version of, that service. In terms of the discussions above, this means that the first partition for defining categories of data to be collected should be baseline/target, where it is understood that "baseline" denotes a known and familiar version of the service and "target" refers to the version of the service that is to be evaluated. This partition effectively doubles the overall sample size. However, experience shows that if a baseline is not acquired in the data collection effort, the measurements produced may not be interpretable.

3.2.3   Data Organization

Once the questions of what data elements and sample sizes have been answered, the next major step in data handling is to begin to acquire the data and to organize the samples into coherent databases from which specific items can be readily retrieved. For small samples, effective data organization of this kind can be as simple as producing a data collection form for recording results and developing a scheme for sorting the forms by date, alphabetical order of site names, etc. This facilitates search through the stacks of forms for particular items or types of items, or recording the data collected into tally sheets. For larger, more complex databases the data should be organized into files on some computer somewhere, so that the quantifiers for the measures chosen for the analysis can be readily calculated, recalculated, and manipulated in the course of trying to answer specific questions.

DBMS or not to DBMS? That is the question

In today's computer literate society the immediate proposal for taking care of data organization will probably be to throw the data elements that are collected into whatever "gee whiz" relational database management system (DBMS) that is currently in vogue with the resident computer professionals. If you get nothing else out of this book, its cost will be rewarded at least a thousand fold if you will understand and heed the following advice as to the wisdom of doing this:

> **Don't do it! Don't even think about doing it! Don't listen to anyone who tries to get you to think about doing it! Don't even listen to anyone who tries to get you to think about whether you should think about doing it!**

Now, don't get me wrong here. For the applications for which they are designed, namely the creation and maintenance of very large databases which have structures that are unlikely to change, such off-the-shelf DBMSs can offer substantial benefits with respect to efficiencies in data storage, speed of responses to queries, and set-up of reports of data summaries that are to be routinely and repetitively generated.

The problem with these DBMSs is that their efficacy is predicated on the user being able to accurately specify the structure of the database in advance of the application of the database construction utility. This means, for example, that the user must be able to support the database design by clearly identifying:

- What different data elements are to be stored in the database and in what order in the records;
- How each element is to be represented (e.g. character, date, numeric value in fields of fixed or variable size);
- The actual content and format of each field, such as category labels and the maximum number of characters for character fields, selection of date formats, and content, maximum values, and floating point precision for numeric fields.

When all of the possible nuances and subtleties of content in the data to be organized are clearly understood and anticipated at the outset, these design specifications work, and the database creation and maintenance with a DBMS can be proceed smoothly. Otherwise, there will be tremendous processing complexities and overheads associated with such revisions as adding new data fields and associating them with their proper records, or expanding fields to accommodate larger numbers or longer character strings than were anticipated.

Worse yet, the effort required to change the semantic content of the data fields for the rigidly structured databases organized by such DBMSs is so daunting that many databases created in this way will be found with obsolete or ambiguous data dictionaries and look-up tables, because it is too difficult to effect the refinements that may be necessary to handle unanticipated needs. To see this, suppose that we are going to set up a database of trouble reports from residential subscribers of a long distance telephone service, with the objective of being able to determine whether complaints are increasing or decreasing for different types of service. Suppose further that one of the major distinctions among services that might affect how often users register a complaint recognized at the outset the database design is presumed to be whether the call was to be billed to the originator or recipient of the call. Since services for which the called party pays are assigned particular area codes, a natural way to distinguish free phone calls from others would be to ensure that the database includes the NPA (area code) for each call in the trouble report database and setting up a look-up table that would identify which NPAs are reserved for free phone services. This way, as other area codes, such as 888, can be added to the look-up table, as they are allocated to freephone service, without any need to change the database structure.

This sounds good and works well until some decision-maker wonders whether the likelihood of users filing complaints on free phone services depends on some other factor, such as whether the call is answered by a person or a voice response unit (VRU). That decision-maker is then not likely to be pleased to be told that the necessary data cannot be produced from the database without the 3 months' effort that will be required to: research the information on all of the different free phone services in use to determine whether a particular NPA/NNX pair is answered by a person, a VRU, or a person or VRU depending on time of day and circumstances; add an NNX field to the database; and create and maintain the huge, frequently changing look-up table required to support queries keyed to the way a free phone service is answered.

Finally, if the possibility of such pitfalls (which accounts for why designers of large databases seem to be constantly involved on documenting and reviewing requirements) is not enough to scare you off, be warned that the auxiliary report generation utilities provided in such systems are tailored for stratified data summaries. Consequently, they will, in general, efficiently handle only the most basic data extraction and manipulation functions, such as filtering against simple criteria and calculation of common statistics, such as averages and standard deviations. Anything more complex can become an absolute script-writing nightmare.

If you have any doubts as to the truth of this assertion, go find someone familiar with SQL to tell you what would be involved in writing a script to process two field records of events comprising an event type designation and an associated field containing a start time or a stop time, when it is known that no new event of any type will start until the previous one has stopped.

APL! How do I love thee? Let me count the ways

These observations suggest and bitter experience proves, then, that off-the-shelf database management systems may represent an attractive convenience early on in the effort to organize data for an analysis, but are in the long run so constraining that the early convenience can be expected to be rewarded by downstream difficulties that will be very costly, and may, in fact, make it prohibitively expensive to achieve the desired utility.

The alternative is to organize the acquired data in a way that will afford the same ease of database creation as is afforded by the capabilities to extract data sets from a rigidly structured database created with a DBMS. The motto for doing this is:

ARCHIVE ON RAW AND QUERY BY EXTRACT

which succinctly expresses the idea that the best way to organize data for purposes of analysis is to:

- Store the raw data in a form that preserves as nearly as possible its original content and context, including any data elements that have no apparent utility in the efforts at hand, but can be readily acquired along with the necessary data elements; and
- Use a general-purpose computer language to create routines that can parse and filter the raw data as necessary to produce a working database tailored for a particular objective.

The early inconvenience in this approach is that the analyst or a support programmer must become proficient in writing routines that will reliably grind through piles of ugly, sometimes ill-formed objects to create a coherent collection of records containing exactly what is needed to answer a specific question. The later pay-off is that any changes in required data elements, definitions of categories, criteria for eliminating spurious records, etc. can be accomplished as readily as, and with much more ease and flexibility than, a DBMS user can write query language scripts. Moreover, the tailored databases extracted via computer routines will then be well-defined objects in the computer language that created them, so there are no import/export/translation activities required to enable their manipulation.

Now, this advice on how to approach data organization would be nothing more than some sort of idealistic theory without a practical basis, except that there is at least one computer language that is ideally suited for this application, and the efficacy of implementing capabilities to archive on raw and query by extract with this language has been demonstrated in literally hundreds of database organization applications, including many in which I have been personally involved. The language is APL (*A* *P*rogramming *L*anguage), which was first described by Kenneth E. Iverson in 1963, first implemented as an interactive programming capability on IBM 360s in 1966, an subsequently very nicely transported to PCs by STSC/Manuguistics/APL2000 (see *www.apl2000.com*) beginning in the early 1980s.

In his book *Applications Development Without Programmers*, James Martin, the 1980s author without peer of books on computing and data telecommunications, observed that database organization and analysis capabilities could be expected to be developed with APL about 30 times faster than their implementation in a compiled language like FORTRAN, making APL the undisputed language of choice for rapid applications development. And, 35+ years of advances in computer science and technology have yet to produce a serious challenger to its position.

What makes APL particularly well suited for DBMS-like applications is that:

1. As a data structure, any relational database can be visualized as a matrix whose rows are records of data elements and whose columns represent fields in each record.
2. APL is nonpareil as a language for defining and manipulating matrices by adding or deleting rows, adding or deleting columns, ordering rows by column values, etc. because it automatically tests the contents of the matrix to determine the variable type for each column and allocate storage space based on the number of rows, number of columns, and the largest and most precise numerical value in each row.
3. APL supports ready specification and execution of searches of matrices to extract rows whose column values satisfy any condition that can be defined by a Boolean expression describing values and/or relationships among the contents of any row in the matrix.

This means, in essence, that APL is a general-purpose computer language that has embedded a very robust relational DBMS that just happens not to be christened with the name. Add to this the numerous other characteristics of the language that facilitate applications development about which, given the slightest provocation, we of the cult of APL-philes will wax eloquent for hours, and you have exactly what you need to be able to archive on raw and query by extract...

3.2.4   Data Manipulation

The fun, and occasionally useful, thing about the creation of computer accessible databases as just described is that it removes barriers to transforming the data to facilitate its understanding. Procedures that would be prohibitively tedious and time-consuming if we had to execute them manually can be executed in seconds, affording the analyst the luxury of exploration data by visualizing, shaping, and fitting the contents of the database. The tools for such open-ended examination of data are referred to here collectively as data manipulation facilities. They include, for example:

- *Visualization aids*, which transform sets of measurements or data elements into graphical displays, such as scatter diagrams, histograms, or time series charts. The choice of which visualizations are most useful for a particular purpose is largely a matter of individual taste and perspective. One of the visualization aids that I find most useful is the cumulative distribution function for a data set, which is a plot that shows possible values for data elements or measures on the x-axis, and the proportion of the total sample whose values are less than or equal to the x-axis value on the y-axis.
- *Calculation of distribution statistics*, which transform the data sets into numbers that generally describe how the data are distributed, such as the mean and standard deviation. Such statistics are convenient, compact descriptors of some of the characteristics of a set of values. However, casual users of statistical calculation facilities should always keep in mind that statistical parameters calculated from data accurately convey some sense of how the values vary only when the values in the data set are normally distributed about the mean, so that, for example, it is equally likely for an observed value to be the same amount greater and less than the mean, and the likelihood of differences of a particular magnitude decreases steadily and goes to zero as the magnitude increases. If this condition is not satisfied, the standard statistics are not very useful descriptors of the data, and may, in fact, be misleading.
- *Data fitting*, comprising utilities to produce the best fit of data sets to a "smooth" mathematical function. The most familiar of such data manipulation capabilities is the use of linear regression against data comprising x,y pairs of observations to produce the best estimate of a and b in the equation $y = ax + b$. Commonly available data fitting capabilities also

include: algorithms for fitting distributions of data with common probability density or distribution functions for well known distributions such as the Weibull, exponential, gamma, and normal; and varieties of polynomial curve fitting routines. The important thing to remember about data fitting is that it has two disparate uses, which must be carefully distinguished. As a data manipulation capability, data fitting is used to extend visualization aids, to *assist* the analyst in determining whether a particular function fit to the data might be accurate enough to support the other use of data fitting, which is interpretation of the data. When data fitting is used as an extension of visualization aids, it should be remembered that the fit to the data displayed is an aid, and not a result. Thus, for example, when the graphical package adds a "trend line" with a positive slope to a chart showing the variations of values of a measure over time, the mere presence of that line does not imply that the value of the measure is increasing. Rather, what is shown is the best linear fit to the values of the measure as a function of time, and that fit may be found on inspection, or by virtue of further analysis, to be so bad as not to imply anything at all about how the values are changing over time.

- *Data filtering*, which is a process by which entire data sets are transformed by eliminating suspect, clearly erroneous, or useless elements. The objective is to "clean up" data sets in order to make sure that all values are free of errors in data acquisition and represent what they are supposed to represent. Elimination of errors in data acquisition might, for example, be reasonably and unarguably based on dropping values in a data set that are outside the operating range of the systems being tested or the range of values that can be produced by the test device. Similarly, it is reasonable to eliminate from a set of outcomes of call attempts placed to a prearranged destination all of the attempts that resulted in a station busy signal, because there is no way to determine whether the busy signal meant that the call was completed to the destination. It might have been that another test was in progress, or the call was misrouted to some other station that just happened to be busy. In order to be sure that a set of values purported to represent measurements of time required to set up a call to a distant station is accurate, it is reasonable to eliminate all connect times that are substantially shorter than the minimum time that can be expected on the basis of known system performance, because these will in all probability represent instances in which the call attempt was diverted to a local treatment rather than routed through the system. Any data filtering beyond that which can be similarly based on sound, concrete reasons, however, is very dangerous, and apt to be abused. For example, one of the classical data filtering techniques is the application of statistical tests to throw out "outliers". What often results from blind application of this technique is an altogether inappropriate exclusion of data points that are truly representative of the process that produced them. Worse yet, the underlying reason for trying to trim the outliers from the data in the first place is frequently found to be that a small number of data points were biasing the average, making its value much greater (or less) than the preponderance of the observations. In cases like these, the only thing that data filtering accomplishes is the masking of what might be very important information embodied in the data set, production of unrealistic estimates of the statistics for the data, and perpetuation of an ill-founded focus on trying to rely on the average as the principal descriptor of the contents of a data set.

As seen from the examples here, data manipulation capabilities must be used with care as repeated use of the same facilities can be addictive and lead to abuse. The principal caution with respect to use of data manipulation capabilities, however, is that whatever they produce is still just data, and data, no matter how well synthesized, is just raw material. A decision-maker may sometimes accept a particular representation of the data as an answer, because the result can be interpreted without further assistance from the analyst. However, none of the displays or descriptors of data produced from data manipulation, no matter how precise, accurate, beautifully composed, or mathematically elegant, can be considered to represent the *information* that is the desired end product of analysis.

3.3 Phase 3: Evaluation

To get to the point of producing information, whatever is derived from the first two phases of the analysis must be *interpreted* to produce answers to specific questions posed by decision makers. The process by which this is accomplished is what has been described here as *evaluation*. The results of such evaluation of measurements will generally address questions related to the incidence or occurrence of undesirable conditions, outcomes, or events by describing the likelihood of their occurrence.

The undesirable conditions may be described in:

- Subjective, qualitative terms, such as: "How often will we experience *outages that will severely inconvenience the people using our business telephone service?*", or "Will the time I have to wait to have a call connected be *annoying?*", or
- Equivalent expressions describing concrete *examples* of unacceptable quality, such as: "How often will we experience *outages of an hour or more?*" or "Will the post-dial delay in the new service be *less than 6 s?*"

As suggested by the use of the word "examples" rather that "criteria" here, any measures and values used in questions at this level will merely represent an attempt to describe concretely and objectively something that is inescapably abstract and subjective, and not an attempt to specify a requirement. For example, a user will know from experience the conditions under which a long PDD will be "annoying", and a user representative will know from past user reactions to outages when an outage will begin to "severely inconvenience use of the telephone". Rather than try to find the words to describe those conditions in concrete terms, or laboriously detail the experiences that give them meaning, the persons raising the questions will frequently resort to subjective estimates of measurable characteristics to exemplify those conditions. In doing so, they will talk about PDDs being less than 6 s or outages being an hour or less, while fully recognizing that neither a PDD of 5.9 s nor an outage lasting 59 min is necessarily acceptable, and neither a PDD of 6.01 s nor an outage lasting 1 h and 15 s is automatically unacceptable.

In other words, even the users recognize that when a particular value of a measure is used to describe an unacceptable condition, there is at best a high correlation between the value of the measure cited and occurrence of the undesirable conditions. Consequently, the answers to questions as to the occurrence of a particular unacceptable event or condition will not be credible unless there is an associated description of the likelihood of experiencing the conditions of concern.

The likelihood may be conveyed in terms of: numerical probabilities, which will usually describe the probability of *not* experiencing the undesirable condition; "fuzzy" descriptions of likelihood of experiencing the condition, such as "highly unlikely", "rare", and "not frequently enough to make a difference in perceived quality of service"; or, occasionally, familiar analogues, such as "it is about as likely as your losing a poker hand when you're holding four aces".

If all of this begins to sound like we have gone back through the looking glass into a very unfamiliar world, it is because the notions of evaluation presented here are predicated on the ideas that:

1. All decisions are ultimately based on assessment of qualities, rather than precisely measured quantities; and
2. The only meaningful quantitative description of qualitatively described conditions is the probability of their realization.

In other words, as a user, I do not decide to buy a particular service because it has an availability of 99.9%; I buy it because the value of 99.9% can be shown to ensure a low probability of experiencing a condition that I want to avoid. Nor do I determine that a particular service is of poor quality with respect to a particular characteristic on the basis of one bad experience; rather, I decide on the basis of how often that experience will be repeated.

Unthinkable isn't it? That all of the machinery that is assembled and applied for hard, objective, quantitative analyses of something like quality of service must ultimately be used to produce information that is expressed in such imprecise, qualitative terms as "good enough", "no noticeable difference", and "unlikely to affect user perception of quality"? However, I assert that it is information expressed like this that is what is actually used in decision-making. It would be very hard to prove this assertion, but it does become much more plausible in light of some very mundane examples of decision-making processes:

- A medical doctor, concerned with diagnosing the health of a patient has the patient's temperature taken. The thermometer reading that comes back is a temperature of 103.2°F. The information of diagnostic value gleaned is that the patient has what can be classified as a "very high fever", which narrows the diagnosis and suggests the need for immediate administration of fever reducing medicines. The specific temperature recorded provides no more information for diagnosis that one of 103.1°, 104° 103.8°, etc.

- A carpenter needs a piece of wood that is 44″ long. He measures one of three pieces available at 43 7/8″, then another at 43 15/16″, and finally one at 44 1/4″. The first two pieces are rejected because the measurement shows them to be "too short", while the third, "can be cut down to the size needed". Another carpenter needing a board of the same length asks his apprentice which of the three pieces can be used. The apprentice does the measurement and returns with the information that "two of the pieces are too short, but you can cut a 44″ piece out of this one". A third carpenter hands his apprentice a piece of wood that is 44″ long and asks whether there is a piece among the three that can be cut to match it. Without ever measuring, the apprentice tests each of the three pieces against the piece from the carpenter and returns with the longest one. Despite the differences in the available measurement data and the ways that the measurements were made, there is no difference in the information that was used to select the 44 1/4″ board.

- A shopper for an item finds it advertised for sale in one place for $15.75 and another, equally convenient location for $16.99. The decision as to where to buy it will not be based on the data that it "costs $1.12 more" at the one place, but that it is "cheaper" there. Another shopper already in a store sees the item for $16.99 and knows that it can be purchased for $15.75 by traveling to the other store. The basis for the decision of where to buy it will not be the data that one can save $1.12, but whether the amount is "enough to be worth the bother" of doing so. When the two prices are $16.98 and $16.99, the difference in price will not even figure into the decision of where to buy the item, because the price is "essentially the same" at either store.

In all these examples, where quantitative data are available, the decision is based on a translation of that data into a qualitative classification, comparison, or assessment representing an evaluation of the data against a particular objective. What might be reasonably quantified, however, is a likelihood of realizing a condition. In the case of the doctor's diagnosis, for example, the utility of the fact that the patient has a high fever comes from a background of knowledge from which the doctor knows that it is highly unlikely for a patient to have a high fever without the presence of some pathological condition. In the case of the carpenters, the various measurement efforts in effect assigned a probability of 1.0 to the condition that one of the pieces of wood could be cut to $44''$, so that the carpenter did not have to go out for more wood. And, in the case of the shoppers, it is clear that one of the considerations that will affect the decision of where to buy the item is whether it is still available at the store advertising the favorable price, given that the difference in price is "worth the bother".

Like everything else in this book, all of this obscure theory has a very practical application to the problem of measurement and evaluation of QoS. In this case, the perspectives support the useful characterization of the nature of the interpretations necessary to convert data into information needed for different types of evaluation illustrated in Figure 3.2. In this figure, the x-axes on the two graphs represent values of a particular measure of a performance characteristic of a telephone call, arranged so that values farther to the right represent worse quality. To fix ideas, suppose that the x-axis represents measurement of the loudness of noise in a telephone call.

Then the top graph, representing the user perspective, displays the probability that a user will find a call to be unsatisfactory as a function of the measured value of noise. As seen there, the user perception is totally unaffected for a large range of values, for which the noise is barely perceptible to the human ear. However, for larger and larger values of the noise level, proportionally fewer of the calls will be found to be satisfactory, until you reach the point that every call with that value or greater will be found to be unsatisfactory.

The likelihood that users will find a particular call unsatisfactory with respect to noise on the line will therefore be determined by the shape of the top curve and the shape of the second curve, which represents the distribution of values of noise measurements for a particular service. Under the assumption that the evaluation of perceived quality of service with respect to noise will require an answer to the question "What is the likelihood that a user will find the noise on a call objectionable?", it will be necessary to use both curves for a particular service to assign a value to the likelihood.

For purposes of evaluating the intrinsic quality of service or monitoring the

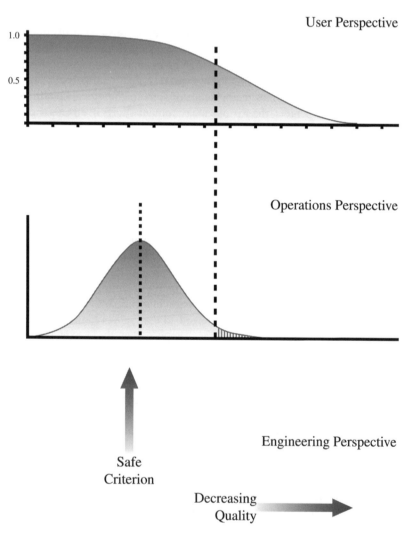

Figure 3.2   Differences in perspectives that determine the basis for evaluation

measurements of noise for indications of deterioration of intrinsic quality, the service operator will be concerned with the shaded area of the second curve, which represents the likelihood that the noise levels in the service will correspond to levels that users begin to find objectionable. The basis for evaluation of the intrinsic QoS is therefore the value of noise measurement shown on the dotted line and the area of the shaded region under the curve. When that area can be reasonably associated with the mean of the distribution of the noise measurements as shown in Figure 3.2, the operations personnel can use the average value of noise over particular circuits as the basis for evaluation.

From the perspective of engineering a service that will be satisfactory to a user community, the system designer will establish a design criterion by selecting a value for the average noise level to be achieved in the new service that is well below that which the operations manager tries to maintain in the existing service.

The cascade of questions that must be answered during the evaluation phase of the analysis of QoS with respect to noise thus looks something like this:

- What is the likelihood that a user will find a noise level of x to be objectionable?
- What average value of noise readings for my service will assure that less that $y\%$ of the calls will be found to be objectionable?
- What average value of expected noise levels should I design in order to be sure that the percentage of calls for which users will experience objectionable noise levels is substantially less than it is in today's service?

At all three levels there is a subjectively described condition (objectionable noise on the line) and a description of its likelihood that is appropriate for the purposes of the audience for the analysis.

Such, I claim, probably is, and certainly can be, the nature of all evaluations of QoS...

4

Telecommunications Concepts

The last step in laying the framework for measurement and evaluation of telecommunications services is to define the terms that will be used in Part II to describe:

- Systems and processes that implement those services;
- Different types of services that must be distinguished in defining audience concerns; and
- Basic user concerns with quality of telecommunications services.

4.1   Basic Systems and Processes

In simplest terms, the basic function of any telecommunications service is to provide a means of electronically exchanging information between remote points. The system that enables such exchanges can be thought of as comprising:

- *Nodes*, which are physical locations of equipment that is used to implement the service;
- *Links*, which are paths between nodes over which information is relayed; and
- *Connections*, a series of node-to-node links connecting the two remote points between which information is to be exchanged.

Telecommunications services, then, provide the means for setting up connections. The node at which the attempt to set up a particular connection is initiated is commonly referred to as the *origin*, while the remote end of that connection is referred to as the *destination*. Origins and destinations for connections may be further described by adding a name describing the type of equipment or location for the connection. Thus, the origin of a connection may be described as being an origin station (or number), an origin PBX, or an origin switch. An *end-to-end* connection, however, is understood to be the complete connection from the first piece of telecommunications equipment that is used at an origin node to the last piece of telecommunications equipment used at the destination node.

To create capabilities to establish connections there will be a set of nodes and links organized into *networks*, which provide standing capabilities to effect connections among geographically dispersed locations. When such networks are designed and operated to establish connections between far distant points, they are referred to as *long-distance* or *wide-area* networks; when the networks are set up to provide connections in a smaller region, they are referred to as *local service* or *local-area* networks.

In order to set up an end-to-end long distance connection, then, the origin node must be linked into the long distance network, the connection must be linked node-to-node across the long distance network, and the connection must be completed by linking a node of the long distance network to the destination node. There are, therefore, three kinds of links and linking facilities that may be distinguished:

- *Access,* by which links are established between the origin node and the first node in the long distance network;
- *Termination*, by which links are established between the last node in the long distance network and the destination node; and
- *Transport*, by which a connection is established between the node in the long distance network linked to the origin site and that linked to the destination site.

Construction and operation of these networks requires two kinds of facilities:

- *Transmission systems*, which effect the node-to-node transfer of information across the network; and
- *Interconnect systems,* which take links coming into a node and connect them to links out to other nodes.

The transmission systems may be set up to carry the information on a single telephone connection, or they may transmit information between nodes fast enough to transport the information being exchanged over a very large number of connections. The oldest, and most familiar kind of interconnect system is the *circuit switch* which receives incoming transmissions, breaks them down into the individual connections to be established, and routes each connection to an outgoing transmission system that establishes a link to the next node in the connection. In today's world, however, there are other kinds of interconnect systems in use, such as digital cross connects, which interconnect two transmission systems at a node without breaking out the individual connections being carried, and *store-and-forward switches*, which take information in from an origin node, and later set up a connection for its transmission onward to the destination node.

Use of these basic networking capabilities to set up an exchange of information over end-to-end connections is supported and controlled by three other processes:

- *Injection/extraction*, used to get signals onto and off the network;
- *Encoding*, used to put signals into the form required for transmission; and
- *Routing*, needed to determine the node-to-node links that will effect a desired connection.

4.1.1   Injection/Extraction

However it is configured, the first step in getting into a network is the transformation of the information into the format required by transmission system(s) that are used for access and termination links. At either end of a connection this is accomplished by *injection/extraction* processing. For simple analog voice transmission, for example, the injection processor is the mouthpiece of the handset, which contains a microphone that transforms the sound waves generated by speech into an electrical signal whose variations in amplitude replicate the variations in force of sound waves against the microphone. The extraction processor is the earpiece of the handset, which converts incoming electrical signals into audible sound waves. A more complicated injection/extraction process is accomplished by a FAX machine, which takes information in its native form as written or printed matter and converts it into electrical signals that are modulated to produce particular waveforms that encode a digital representation of the visual material.

4.1.2   Encoding

Once the information to be exchanged is injected in a form that can be handled by one transmission system, other transformations may be necessary to accommodate different transmission systems that may be used in setting up a connection. These include:

- *Analog/digital (A/D)* and *digital/analog (D/A)* conversions, whereby electrical signals representing waveforms are converted to electrical signals representing strings of bits, and vice versa. The principal A/D and D/A conversions for electrical waveforms are accomplished with devices called *CODECs*, the most common of which employ either some form of *pulse-code modulation (PCM)* whereby an analog waveform is represented digitally by sampling and digitally encoding the amplitude of the waveform at fixed intervals, or *code-excited linear predictive coding (CELP)* whereby segments of analog signals are processed to determine the best fit to a library of segments and the digitally encoded symbol for each segment is transmitted to the distant end, where each segment is re-constructed according to its description in the transmitted symbol.
- *Digital/analog* and *analog/digital* conversions, whereby information injected in a digital format is converted into electrical signals representing waveforms for transmission, and extraction comprises re-creation of the digital bit stream. These conversions are accomplished by devices called *modems (modulator/demodulators)*, which are nowadays readily recognized as the things that make the strange noises when a FAX unit or computer begins to connect through a telephone line.
- *Multiplexing*, whereby different transmission signals are packed together for transmission over links whose transmission system supports throughput greater than required for the individual signals. For RF (radio frequency) carriers of electrical waveforms, for example, such multiplexing is accomplished by measures such as frequency- and time-division, whereby narrow band signals are assigned frequency slots in a broader band signal and/or are chopped into segments that are interlaced with segments from other signals over a faster transmission system. For digital transmission systems, such multiplexing is accomplished by taking bit streams being transmitted at slower data rates and assigning them locations in the bit stream carried by a system with a faster data rate.
- *Data framing*, whereby other bits are added to blocks of data to be transmitted, to format the data for multiplexing, detect and correct errors, control transmission, or describe the contents of the data in the block. When data framing is applied to a block of data, the added bits are referred to as *overhead* and the original block of data is called the *payload.*

4.1.3   Routing

The last important capability required to use a network is the ability to define and control the node-to-node linkages that set up a desired origin/destination connection. This process is referred to generically as *routing*. It is accomplished by use of a combination of:

- *Signaling systems*, which provide the means of communicating the information needed to set up a connection; and
- *Switching systems*, which react to information describing the origin and destination of a desired connection and select the specific node-to-node links that will be used to set it up.

4.1.4   Signaling Systems

Signaling systems used in telephony are characterized as being either *in-band* or *out-of-band*. When in-band signaling is used, the information needed to set up a connection is communicated node-to-node on the same links that will carry the desired exchanges of information. When out-of-band signaling is used, the information needed to set up a connection is communicated node-to-node through a parallel telecommunications system whose transmission systems carry only the information needed to select and establish the node-to-node links used in a connection.

In today's ordinary telephone services, most access and termination routing is accomplished in-band by exchange of tones and other electrical signals over the same lines that will be used for the end-to-end connection, while routing into and across transport networks is accomplished via out-of-band signaling. Thus, for example, a person at station A who wants to place a telephone call picks up the handset, thereby initiating a change voltage on the line, which indicates the desire to place a call. That signal is answered by a steady tone, which is understood to be a dial tone, indicating that the number to be called can be dialed. In response the user (or the user's autodialer) generates a set of dual tone message frequency (DTMF) tones that are recognized down the line as a number. When the DTMF tones are *registered,* i.e. received and recorded into computer memory at a distant switch, the associated number and the number of the originating station are passed into an out-of-band signaling system, which uses that information to query distant switches as to the availability of links that might be used, select a node-to-node route for the connection, and transmit all of the control messages needed to set up the necessary interconnect at each node. At the distant end, the seizure and test of the line to the desired destination station and the signaling back to the origin of the status of a connection (ringing, waiting for an answer; station is busy; connection could not be set-up) are effected by in-band signaling.

Data communications services in which the information to be exchanged is injected as data and is not converted to analog waveforms along the way, generally use in-band signaling, but in a different form. The origin/destination information needed to determine how to get a data block to its intended destination is included in the data framing or the multiplexing routines, so that interconnects can be selected independently and dynamically at each node as the necessary connection is being set up.

4.1.5 Switching Systems

The interconnects that are necessary to set up the node-to-node links in a connection are effected by *switching systems*, which are in effect computers that receive the information as to the destination of a desired connection, examine current information about what links are available for use, and select and specify the interconnect to be used to set up the next link in the desired connection. There are essentially two kinds of switching systems in use today. The first, and most familiar, is *circuit switching*, whereby the node-to-node links in an origin/destination connection are set up via the interconnects, and the connection stays up for exclusive use of exchanges of information between the origin and destination until it is torn down.

However, both the newest and perhaps the oldest switching systems use a different technique for establishing connections between the origin and destination. The technique is *store-and-forward relay*, whereby information is exchanged by transmitting an information element from one node to the other in a process via which: (1) the element is received completely by each node in a connection in turn; (2) the routing information is examined; and (3) the receiving node determines to which node, on which link, and when the information element will next be transmitted. In the ancient version of this kind of switching, the information elements transmitted were complete messages, then called telegrams. These days, however, the same kinds of information elements are called e-mail/electronic mail messages. When the information elements handled in a store-and-forward relay system comprise very small segments of messages or conversations, those elements are called *packets,* and the technique is in this case referred to as *packet switching*.

4.1.6 Types of Service

To help fix ideas, Table 4.1 provides examples of the various kinds of systems and processes that were just described. However, specific systems are of little interest here. Rather, the principal objective in going through those descriptions was to be able to distinguish and discuss in general terms different types of telecommunications services, whose use may generate different concerns. Specifically, what I have in mind here is distinctions among the types of

Table 4.1   Examples of elements of telecommunications systems

Transmission		Analog telephone subscriber loop on copper wire
		DS0 64 kbps carriers on copper wire
		DS1/E1/T1 carriers on optical fiber, microwave or satellite RF paths
		SONET/ATM optical fiber transmission systems
Interconnect		Digital cross-connect matrices/"soft" switches
		Patch panels
		DMS250, ESS5, AXE circuit switches
		PBXs
Injection/ extraction		Handset microphones
		Telegraph keys, teletype tape readers and printers
		Channel banks
		Data and FAX modems
Encoding	A/D, D/A	Voice codecs
	D/A, A/D	Systems implementing ITU V.x coding and decoding
	Multiplex	DSx/DSy mutiplexers for digital transmission
		TDMA, FDMA and CDMA mutiplexers
		GSM bandwidth sharing systems
	Data framing	T1/E1 framing and superframe generators
		ATM/SONET virtual channel specification and identification subsystems
		IP packet encoding subsystems
Routing	Signaling	*In-band*: DTMF/dial pulse transmission and reception systems
		Out-of-band: Systems implementing ITU CCS7 routing protocols
Switching		*Circuit*: DMS250, ESS5, AXE circuit switches/PBXs
		Store-and-forward: e-mail/telegraph relay centers
		Packet: IP/ATM packet switch routers

connections from origin to destination, the way they are set up, and the way that the service over the connection is priced and billed, all of which may affect user expectations and concerns.

4.1.7 Types of Connections

Types of telecommunications connections are defined by what is injected and extracted over the connection. When the input to the injection processor at the origin is sound waves, the type of connection established is a *voice* connection, even though the waveforms may be digitized, as happens with most cellular telephones and some modern digital handsets. When the injected content is digital data, such as computer data files, the connection is characterized as being a *data* connection. In a data connection, the output from the injection processor at the origin to the transmission system may be a bit stream that can be directly multiplexed onto a digital transmission link, or it may be waveforms that represent patterns of 0s and 1s created by modems. To distinguish these two possibilities whenever necessary to avoid confusion, the former type of connection will referred to as a *direct data* connection and the latter will be called an *acoustic data* connection.

Since voice signals can be digitized, voice can be readily transmitted as digital data, as long as the voice data frames can be transmitted fast enough to maintain continuous regeneration of the waveforms needed as inputs to the distant extraction processor. Consequently, there is a potential for connections, such as those effected by ISDN, that can support either a voice or data or both at the same time. Such connections will be referred to as *hybrid* connections.

4.1.8 Set Up

In general, node-to-node connections into, out of, and across networks can be thought of as being:

- *Dedicated*, meaning that the connection, once set up, is left up indefinitely, so that it is always active and ready for use as needed;
- *Circuit-switched*, so that the connection is set up on call as a series of node-to-node links and left up only as long as the origin/destination connection is required;
- *Packet-switched*, so that the capabilities to effect transfer of small segments of data blocks, messages, or conversations are always in place and transmission is handled via store-and-forward relay among the nodes in the packet-switched network;

- *Message-switched*, so that capabilities to effect transfer of complete messages are always in place and transmission is handled via store-and-forward relay; or
- *Hybrid*, so that part of the node-to-node connections are set up with one technique, and part are set up with another, as happens, for example, when a long distance connection is circuit-switched to a packet-switched network for transport and then circuit-switched at the other end to connect the call at the terminating long distance switch.

The end-to-end set up of a connection between an origin and destination may be accomplished by use of any of these set up techniques for access, transport, or termination. Thus, for example, a large facility may have dedicated access to a circuit-switched long distance transport network, so that every call attempt draws dial tone from a switch in the long distance network. Connections to the destination may then be effected via dedicated terminations from the long distance network to the other large facilities, or via terminations that are circuit-switched through the local service network.

This implies that there may be as many as 15 different ways a particular connection can be set up. Many of these combinations are rarely, if ever, used on the other side of the looking glass, so it will not be necessary to consider all of them at every turn in Part II. It is, however, important to understand at this juncture that there are a lot of different ways of setting up end-to-end connections across a given network, and each variation may suggest a different quantifier for a given measure of quality of those connections.

4.1.9   Billing Method

Another factor that will surely color, if not completely shape, user perception of quality of a particular service is the way that it is billed. The principal distinctions in this regard are who pays for the service and how the usage is billed. In the case of who pays, the question is whether: (1) the billing is *regular*, so that the calling party pays, or (2) *free phone*, so that the called party pays for the usage. In the case of how the usage is billed, the question is whether: (1) the service is *metered*, so that billing is based on capacity used, or (2) the service is *leased*, so that billing is based on a fixed number of facilities or a specified capacity, and it does not matter how much that capacity is used.

In these terms, then, most home telephone local services are leased, while long distance calls are metered, and the billing of long distance calls from home is regular, except for those calls placed to 800 numbers or their equivalent.

As should be obvious at this point, the reason for making these distinctions is that the differences among methods of billing defined here can greatly affect both user expectations of, and user satisfaction with, a telecommunications service. For example, users whose local access is metered instead of leased will be a lot more sensitive to post-dial delays, blocked calls, etc. in the long distance network, because they are being billed locally for such worthless connect time, even though there is no long distance charge. Similarly, a user's reaction to being placed in a queue and asked to "wait for the next available agent" will be entirely different when the number called is a free phone number rather than one that is regularly billed.

4.2 Basic User Concerns with Service Quality

As suggested at the outset of this chapter, one of the keys to effective measurement and evaluation of QoS is to base the definition of measures on an examination of the likely concerns of users of the service, so that all analysis can begin with perceived QoS. This admonition leaves open, however, the question of exactly what those user concerns might be. The answer is predicated on Table 4.2, which, in the spirit of Robert Fulghum's *All I Really Need to Know I Learned in Kindergarten*, displays a very simple view of what users expect from their telephone services. Its purpose is to identify and give names to operational characteristics of a telephone service that determine how often the response to each action in the process of using a telephone will conform to user expectations.

Now, if you buy the idea that users' perception of quality of a telecommunications service is based on what they experience, rather than what the system does, then each of the operational characteristics named in Table 4.2 will manifest itself in day-to-day use of the telephone, as users sporadically encounter unexpected responses or conditions. Mere use of the telephone will thus generate user concerns associated with each characteristic. These are expressed in Table 4.3 as questions whose answers will determine what the user thinks of the quality of the service.

Such is the answer to the question as to *what* must be measured in order to evaluate perceived QoS. In addition, the formulation of the user concerns in Table 4.3 immediately suggests three principles that should be applied in deciding *how* to define associated measures:

- *Measures of QoS should clearly relate to users' concerns.* If one begins with an enumeration of user concerns like that shown in Table 4.3, it is easy to see that user concerns are ultimately formulated as doubts and expressed

Table 4.2 Users' expectations that shape their perception of long distance telephone service quality

User's action	Expected response/condition	Characteristic
Takes phone off-hook and dials the access number for long distance	Dial tone from the long distance service	Accessibility
Dials distant station number	Timely indication that the call has been routed	Routing speed
Hears network response	Ring back or station busy signal	Connection reliability
Hears answer from distant station	Answer will be the station/device called, not some other station	Routing reliability
Holds conversation/exchanges data	Readily audible, unimpaired, recognizable speech/signals	Connection quality
	No difficulties hearing/reading by person/device called	
	No premature interruption of the connection	Connection continuity
Concludes information exchange and hangs up	Tear down of the connection and cessation of billing	Disconnection reliability

Table 4.3 User concerns with quality of telephone services

Accessibility	Will I be able to get to the service when I want to use it?
	How long will I have to wait if I can't?
	How often will the wait be really bothersome?
Routing speed	How long does it take before I know that a connection is being set up?
	Is the time predictable?
Connection reliability	When I dial a number will the service set up a connection to the distant station, or let me know when the station is busy?
Routing reliability	If I dial the number correctly, will the service set up the right connection?
Connection quality	How good will the voice from the distant station sound?
	Will I be heard and understood without difficulty?
	When I transmit or receive data, what kind of throughput can I expect?
Connection continuity	Will my voice connection stay up until I hang up?
	Will data exchanges complete without premature disconnection?
Disconnection reliability	Will the connection be taken down as soon as I hang up?
	What happens if it isn't? Is there someone who will believe me when I tell them that I did not talk to my mother-in-law for six solid hours, and correct the billing?

as questions; those doubts, then, are best addressed via measures that directly and specifically answer the users' questions.

- *Measures of QoS should be defined in terms of what users experience.* As suggested by Table 4.2, each of the operational characteristics nominated for measurement here is associated with a well-defined step in the process by which the telephone is used to place calls and hold conversations, and a response or condition expected by users at each of those steps. Although such a partition may seem unnatural when contrasted with the perception of the underlying processes by persons who are familiar with telecommunications technology, it will be readily recognizable by anyone who has used a telephone and therefore be more readily understood. Since quality is in the final analysis a user concern, it is therefore much more logical and useful to define measures of QoS that recognize the processes, activities and outcomes that are naturally apparent to users.

- *Measurement of QoS for any operational characteristic should recognize and reflect any multi-dimensional assessments that might be invoked by users.* This is a subtle point that is illustrated by the formulation of the user concerns in Table 4.3. If we are talking about service accessibility, for example, there are two aspects to the concern – how often the system will be inaccessible, and how long it will take before it is accessible, once access is lost. Consequently, we can readily envision a user deciding that outages are occurring far more frequently than is comfortable, but are, at least, not lasting very long. Similarly, there are two aspects to routing speed – how fast calls are being set up, and how much variation there is in the times, so that we can envision a user determining that the post-dial delay is generally longer that it should be, but can be accommodated, because it is consistent, or that the post-dial delay is very small most of the time, but the sporadic occurrence of very long delays is a major irritant. Such possibilities, together with the possibility that any one of the mutually independent operating characteristics described here may, in turn, be determined by mutually independent system characteristics should figure strongly in the formulation of metrics for QoS.

4.3   Preview

As will be a relief to some and a disappointment to others, this ends the Cheshire Cat dialogues with our trying to understand that which way you ought to go depends on where you want to get to, and Part II turns to the much more down-to-earth problem of actually defining measures and quantifiers for different aspects QoS. In that effort, the development will be keyed to the basic user concerns just defined and the various types of service defined earlier. Each major division will be devoted to problems of measurement and evaluation of one of the basic user concerns named in Table 4.3, and different types of service will be identified according to the classifications defined earlier as necessary to distinguish appropriate differences among possible quantifiers for the measure(s) defined. After addressing measurement and evaluation of perceived quality, the exposition will then turn to definitions of indicators that might be used in lieu of the actual quantifiers for purposes of monitoring operational performance, again tying the development to distinctions of different types of service as necessary to identify possible variations in the indicators.

But, then, why am I telling you this, when you can see it for yourself by turning the page?

Part II

Evaluative Concepts, Measures, and Quantifiers

5

Overview

In my salad days as a communications operations analyst for the US Navy, any diversion into the kinds of analytical perspectives laid out in Part I would invariably prompt someone in the audience to pull me back out of the rabbit hole by saying, "Why are you telling me all this? I've asked you for an orange and you're telling me how to plant an orange tree!" This, indeed, is what I have been doing throughout Part I, although the instruction in this case is more of the order of a course in citrus horticulture, with occasional tips on grove management thrown in for good measure.

But, now it is time to pass out the oranges that I claim are made particularly juicy and flavorful by use of the fertilizer recommended in Part I. To this end, recall that Tables 4.2 and 4.3 displayed in Chapter 4 define seven basic user concerns with quality of telecommunications service – accessibility, routing speed, connection reliability, routing reliability, connection quality, connection continuity, and disconnection reliability. Here each of these is used in turn as the focus for defining measures and quantifiers for quality of service (QoS) of different kinds of telecommunications services. In each case the development:

- Begins with a discussion of user concerns and a definition of generic measures of quality suggested by those concerns;
- Proceeds from the definition of the generic measure to descriptions of quantifiers that are appropriate for various different types of services; and
- Concludes with a discussion of how those quantifiers can be interpreted to evaluate perceived QoS.

In some cases, the development extends to definition and analysis of associated measures and quantifiers of intrinsic QoS. Such excursions are, however, taken only in those instances in which a quantifier of intrinsic QoS is deemed to be particularly useful for purposes of evaluation, because:

1. It can be independently estimated and used in conjunction with other factors to quantify a measure of perceived QoS;
2. It can be used in lieu of quantifiers of perceived QoS to determine likely user comparisons of the quality of competing services; or
3. It can be readily monitored for reliable indications of changes in QoS over time.

Finally, in Chapter 13, there is an open-ended discussion of some of the factors other than perceived QoS that ultimately determine whether a service will be found to be satisfactory, or condemned as something to be replaced "…in less than no time…" at no more than any cost.

6

Accessibility

6.1 Evaluative Concepts

6.1.1 Examples

The first item on the list of user concerns is *accessibility*, which refers generally to concerns expressed by users who are exposed to, and recognize the possibility of, conditions that make it impossible to set up end-to-end connections normally supported through a telecommunications service. Such complete interruptions of service usually occur only when equipment, software, or connecting lines fail in such a way as to take down all possible links serving a node somewhere in the network. Because modern switched transport networks are multiply connected to ensure that connections can be made across the network even when there is a failure of all links between two nodes or complete destruction of a particular node, this means that such service interruptions are usually attributable to failures of dedicated facilities linking user sites into switched transport networks, or to failures of node-to-node links in thinly provisioned private transport networks. The most frequent causes of such linking failures are equipment failures, or severing of the fiber optic or copper lines between nodes. However, they may also be caused by problems with routing software that keep the system from accepting any requests for connection from particular origins or completing any connections to particular destinations.

Illustrations of the possibilities of service interruptions include the following events.

- A carpenter who is building an addition onto your house breaks through an outside wall and manages to cut the telephone line at the point that it comes into the house. No one can call into you, and you can't call out – connection impossible.

- You leave your handset off the cradle and don't hear the warning signal. The local switch temporarily disconnects the open line, so that the phone is "dead" when you pick up the handset to make a call, and remains "dead" for some period after you return the handset to the cradle. During that period you cannot call out. Moreover, anyone who tried to call you any time after the local switch disconnected your open line got a false busy signal, probably wondering what on earth you were talking about for so long.
- Long distance service out of your building is handled via a direct access link to your long distance carrier's switch that is "nailed up" through the local service provider switch. A city waterworks maintenance crew digging for pipes comes upon this strange looking tree root and applies the chainsaw to get it out of the way...
- Your company's wide area network (WAN) travels on a single threaded microwave transport from city A to city B. The area traversed by one of the microwave shots is deluged with a severe rain and hailstorm that drops 5 inches of precipitation in 35 min. The connection between city A and city B is up and down like a yo-yo, suffering seven rain outages lasting between 1 and 3 min over the life of the thunderstorm.
- Because of a clerical error, your mobile telephone registration number is coded 'STOLEN'. All attempts to log in are rejected.

Similar examples of outages that do *not* result in interruption of service are:

- The carpenter cuts through an inside wall, severing the inside phone line. All of the phones in the back half of the house are "dead", but you can still get a dial tone on the phone in the kitchen. You are inconvenienced by having only one place to use the telephone, but there is no interruption in your phone service.
- You have two phone lines into your house, and you leave the handset off the cradle when the first line was selected. Even though line 1 is unusable for a while, you can still originate and terminate calls on line 2 – if you can keep your teenage daughter off it.
- Long distance service out of your building is handled on a direct access link carried on a fiber optic cable that runs directly from your building to the long distance carrier's switch. When the city waterworks crew does its root pruning on that cable, the cable to the near-by local switch is untouched. The worst consequence is that your special price virtual private network service might be interrupted; long distance service is still accessible via 1 + dialing through the local switch. Moreover, if some of the circuits carried

on the cable to the near-by local switch are "nailed up" as direct access lines to the long distance carrier switch, the special price virtual private network service may be overloaded, getting a lot of all trunk busy signals, but it is not interrupted.

- Your company's WAN travels on a SONET ring. The ring segment between city A and city B goes out due to a repeater failure. In about 10 s, the routing is switched to the other direction and the connection between city A and city B is restored the long way around the ring. Since the file transfer protocol on the WAN is set for 25 s time-outs when data flow is interrupted, and short delays exchanging data are normal under the Internet protocols, the 10 s outage between city A and city B produces no perceptible interruption in service.

As for the mobile phone whose registration is tagged 'STOLEN' - afraid not. The hardware is fine and the ability to transmit and receive is intact, but the telephone service is hopelessly interrupted.

These examples suggest, then, two important principles that should be kept in mind when analyzing QoS with respect to accessibility:

1. *Interruptions of service are not the same as outages.* As demonstrated by the second set of examples above, outages on the facilities that support a service are not necessarily manifested to the user as service interruptions. In particular, it should be noted in this regard that: outages in switched transport networks rarely result in service interruptions; failures of some, but not all, of the service access links from a site may be manifested as difficulties with setting up connections, but do not result in service interruptions of the kind reflected in concerns for accessibility; and there can be substantial interruptions in service even when there are no outages of equipment or transmission media.

2. *From the perspective of users, service interruptions affect accessibility only when they are detected and unexpected.* Consider, for example, the case of the temporary disconnection of service due to the hand set being left off the cradle. If a user notices the handset is off, replaces it without checking the line, and does not try to place a call for a while, it makes no difference to the user that the line was temporarily "dead". However, if the condition is detected because the user tries to make a call and finds that the line is "dead" after replacing the hand set and hammering on the cradle switch, then the service interruption is viewed as problem. Similarly, if the city waterworks chainsaw the fiber optic cable late at night, when no one in the building is making long distance calls, and the service provider mends the cable before sunrise, the extended service interruption will go totally unnoticed as the users return for work the next day. Moreover, a service inter-

ruption that is detected will have little effect on user perception of accessibility if it is an expected event. For example, that mobile telephone user whose service is hopelessly interrupted will probably not become aware of the problem until calls are attempted from known good areas, because service interruptions due to gaps in coverage are an expected and accepted characteristic of cellular/PCS phone services. Similarly, if there are outages during a period when the service provider has negotiated a maintenance window and has announced in advance that service is likely to be interrupted for preventive maintenance actions, users will tend to discount any inconvenience experienced. To underscore this principle, where there is any possibility of ambiguity, service interruptions that are both detected and unexpected will be referred to *operational service interruptions*, or more briefly *OSIs*.

6.1.2   *Variations with Type of Service*

When dealing with accessibility, there are two distinctly different types of user concerns, mandating two distinctly different measurement and evaluation schemes. In what follows, the two different schemes are distinguished as applying to services that are *intermittently* or *continuously* used.

6.2   Intermittently Used Services

Intermittently used services are those, like dial-up telephone services, used to set up connections only when users have a need to exchange information. Such services may be implemented with switched or dedicated facilities, but the distinguishing feature is that there will always be times when there is no need for the service interspersed with the times that it is needed and used.

6.2.1   *Concerns*

The user concerns with accessibility of intermittently used services translate to two questions:

- How often will I experience operational service interruptions?
- How soon will service be restored when one occurs?

These two questions will frequently be expressed by users as a single question:

- How often will I experience an operational service interruption lasting x (minutes, hours, days) or more?

in which x represents a duration of an operational service interruption typifying something that is unacceptable to the user.

6.2.2 Generic Measure

The first two questions in the set above show that there are two underlying concerns with accessibility of intermittently used services, both of which must be addressed to reassure users that operational service interruptions will be neither so numerous nor so long as to be intolerable. The generic measure that answers the first question is the *expected frequency* of *OSIs*, and the one that answers the second is their *expected duration*. These have classically been combined into a single ratio to define measures of accessibility by setting:

$$A = 1/(1 + fd) \tag{1}$$

where A denotes an index of accessibility, f denotes a measure of the frequency of occurrence, expressed in some unit of time, and d denotes a commensurate measure of the average duration of service interruptions. For example, when failures, denoted 'F', are understood to be operational service interruptions, and 'R' stands for 'restore', Eq. (1) can be shown to be equivalent to the classical ratio:

$$A = MTBF/(MTBF + MTTR) \tag{2}$$

However, combination of the expected frequency and duration of operational service interruptions into single indices like those shown in Eqs. (1) and (2) suffer from a fatal weakness, in that the combined measure loses the specific information needed to address both concerns. The possible consequences of this loss of information are illustrated in the dialog between an analyst and a marketer recorded in Appendix B in which the analyst tries to explain to a marketeer why availability quantified by such ratios is worse than useless as a metric of perceived QoS.

To preserve the information in one measure necessary to address both user concerns, the generic measure of accessibility recommended and used here is AC[t], the probability distribution for duration of interruptions, defined, for example, with hours as the unit of time as:

AC[t] = the probability that an operational service interruption
            will last t hours or longer

The definition of the measure as a function, rather than a single number, creates a single quantitative description of OSIs that can be interpreted as needed to address any specifically expressed user concern with accessibility of a particular service in the form of the third question above.

6.2.3   Quantifier

For intermittently used services, the most useful quantifier that can be derived from AC[t] is a graph, referred to here as *operating characteristic curve (OCC)* for operational service interruptions. On such a graph the x-axis is time, representing durations of OSIs, and y-axis is mean time between occurrences, so that the value of y for any x represents the expected time between OSIs lasting x time units or longer.

To illustrate the way such an operating characteristic curve is created, Table 6.1 presents a hypothetical chronology for a telephone service comprising a single subscriber line terminating to a single, leased telephone set that is provided and maintained by the service provider, much as existed in nearly all households with a telephone in the 1960s. Table 6.2 then summarizes the data on operational service interruptions derived from that chronology. The top of Table 6.2 shows the raw data from a period of observation of 3653 days, or 87 672 h; at the bottom of Table 6.2 the interruptions are rearranged in ascending order of their durations to produce a table of data points derived from the raw data.

Plotting of the data points in Table 6.2 on a log/log scale then produces the empirical operating characteristic curve for the single line, leased telephone service shown in Figure 6.1. This empirical operating characteristic curve conveniently summarizes the 10-year experience of one particular user of a particular service (who obviously did not have enough else to do to keep occupied), providing a very good idea as to how often one might expect service interruptions of a particular duration. Were that same data collected by a larger body of users over the same period, or an even larger body of users over the period of a year, and processed in the same way, we would expect the OCC to smooth out, perhaps as shown by the dotted line in Figure 6.2. Such a smooth curve derived from a large sample would then form a basis for characterizing the expected performance with respect to operational service interruptions of any duration asserted by a user as typifying an unacceptable condition.

For example, were a user to express a concern as to how often an operational service interruption lasting 1 h or more might be expected for our hypothetical service, then the smoothed curve in Figure 6.2 could be used as shown by the first set of dotted vertical and horizontal lines to suggest such an interruption can be expected to occur once about every 8500 h, or about once a year. At the same time, users who say that service interruptions do not really begin to be "bothersome" unless they last 5 h or more can be reassured that the mean time between occurrences of such bothersome events will be more like 15 000 h, or about once in about 2 years. Neither estimate is a

Table 6.1   Phone service diary 1960–1970

Monday, 4 January 1960	New telephone installed and checked out
Tuesday, 12 January 1960	Tried to call out at 10:00 a.m. No dial tone. Called for repair from neighbor's house at 12:30 p.m. Repairman refastened loose connection at box at 5:30 p.m.
Sunday, 23 October 1960	Returned from a weekend vacation. The neighbors said that while we were gone an automobile had knocked down the telephone pole on our corner, taking out both the power and phone services for about 3 h.
Monday, 14 November 1960	Arose at 6:00 a.m. to discover that an overnight ice storm had snapped the line into the house overnight. Line was repaired on 15 November at about 5:30 p.m.
Thursday, 25 May 1961	Dropped the handset and broke the earpiece about 11:30 a.m. Called repairman about 1:30. Telephone was replaced at 3:35 p.m.
Tuesday, 18 June 1963	Heavy wind and rainstorm. Branch from a tree blew down across our house lines about 1:30 p.m., breaking everything. Power and phone company repair crews on site about 5:30 p.m. Repairs completed about 9:00 p.m.
Sunday, 17 November 1963	Came home and found that the cat had knocked the telephone onto the floor. No dial tone when I put everything back up. Checked back and got dial tone after about 15 min.
Friday, 28 August 1964	At 9:00 a.m. workmen coming in to build a fence backed their truck under the phone line, tearing it out. Repairs made by 1:30 p.m.
Tuesday, 20 July 1965	Got up at 8:00 a.m. and found that flooding overnight had disabled the local area switch serving our exchange. Lines were transferred to an alternate on a temporary basis, and service was restored by 2:00 p.m. on 21 July. Lots of congestion, but we are able to get calls through.
Thursday, 3 February 1966	New electronic switch cut in, giving us push-button dialing capability. Because of a programming error, all our attempts to dial out with tones result in fast busy. First experienced the problem at 8:00 a.m. Problem was corrected by 1:30 p.m.
Monday, 7 February 1966	Started getting fast busy signals again at 8:30 a.m. Problem cleared by 9:00 a.m.

Table 6.1 (*continued*)

Tuesday, 8 February 1966	Same problem with fast busys at 8:00 a.m. Problem cleared by 8:20 a.m.
Thursday, 16 Jun 1966	At 1:00 p.m. keypad on the telephone stops working; punching keys does not send tones. Unit replaced at 4:12 p.m. by a repairman who suggests that punching the keys harder does not make them work better.
Tuesday, 11 April 1967	Phone off-hook again. Replaced at 10:00 a.m. Was able to call out by 10:20 a.m.
Tuesday, 3 December 1968	Tried to make call. Line was "dead" for no apparent reason. Tried again 5 min later and got out OK.
Thursday, 20 November 1969	Another big ice storm. Lines down all over the area. Lines into the house OK, but could not get dial tone between 2:00 p.m. and 3:15 p.m.

guarantee, but either estimate is both more specific and more meaningful than the assertion that the average accessibility, calculated from the data in Table 6.2 is 99.89%, or even that the overall mean time between OSIs is 6272 h, with a mean time to restore of about 7 h.

6.2.4   Availability vs. Accessibility

As it has been defined here, accessibility is a measure of perceived QoS. The corresponding measure of intrinsic QoS is *service availability*, which is defined in much the same way as service accessibility, but from the viewpoint of the service provider rather than the service user. The underlying concerns are driven by the same recognition that there are conditions that may totally interrupt service, but expressed in terms of *outages*, rather than what users perceive as the result of those outages.

The provider's concerns with availability of services thus translate to two questions that are the analogs of the ones expressing user concerns with accessibility:

What is the expected frequency of outages that completely interrupt service?
How quickly can we restore service when such an outage occurs?

The difference here can be readily understood by examining Table 6.1 for differences between the durations of the perceived service interruptions and the underlying outages that caused them. Such an exercise will readily show that:

Table 6.2   Data derived from Table 6.1

Service interruptions (h):		
12 Jan 1960	7.5	
14 Nov 1960	35.5	
25 May 1961	4.08	
18 Jun 1963	7.5	
17 Nov 1963	0.25	
28 Aug 1964	4.5	
20 Jul 1965	30.0	
3 Feb 1966	5.5	
7 Feb 1966	0.5	
8 Feb 1966	0.33	
16 Jun 1966	3.2	
11 Apr 1967	0.33	
2 Dec 1968	0.08	
20 Nov 1969	1.25	
Number: 14	Total: 100.52	
Period of observation:	3653 days = 87672 h	

Service Interruption Duration x = (h)	Number with duration of x h or more	Mean time between interruptions (= 87672/number)
0.08	14	6272
0.25	13	6744
0.33	12	7306
0.50	10	8767
1.25	9	9741
3.20	8	10959
4.08	7	12525
4.50	6	14612
5.50	5	17534
7.50	4	21918
30.0	2	43836
35.5	1	87672

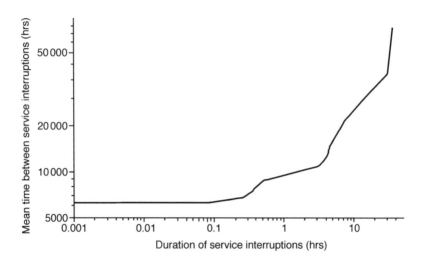

Figure 6.1 Operating characteristic curve for operational service interruptions on a hypothetical single line, leased instrument service

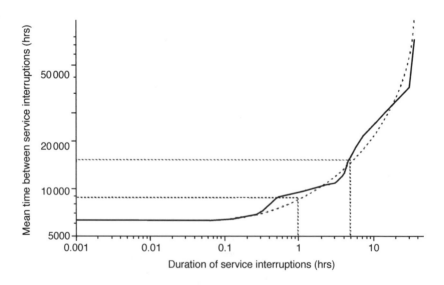

Figure 6.2 Operating characteristic curve for operational service interruptions on a hypothetical single line, leased instrument service, showing expected smoothing

1. The duration of the service outage and service interruption can be the same. (see, for example, the entry on Friday, 28 August 1964)
2. The duration of the service outage may be much longer than the perceived service interruption. (Monday, 14 November 1960 and Sunday, 17 November 1963)
3. The perceived duration of the service interruption may be longer than the actual service outage. (Monday, 7 February and Tuesday, 8 February 1966, since the correction may have been made as early as 8:05 a.m., but the user did not confirm the correction until later)
4. A service outage may not result in an operational service interruption at all. (Sunday, 23 October 1960)

By exploiting the similarities of concerns and accounting for possible differences, any quantifier for service availability can be transformed into a quantifier of service accessibility. This is accomplished by producing estimates of:

$PP[x]=$ the probability that a service outage of duration x or longer

will be perceived as an operational service interruption, and

$\Delta[SO, SI|ty] =$ the average difference between duration of service

outages and duration perceived service interruptions

as a function of ty, denoting a type of outage,

and using these to effect appropriate adjustments to the quantifier for service availability.

In the case of intermittently used services, for example, the desired quantifier for accessibility is an OCC for operational service interruptions. When there is no readily available source of data on frequency and duration of OSIs like that shown in Table 6.1, the desired OCC can be produced from data on service availability and the estimates defined above by applying the following transforms.

MTBOSI *from* MTBO *and* PP[x]

As in the case of accessibility, the most useful generic measure of service availability, AV, for intermittently used services is the probability distribution:

$AV[t] =$ the probability that a service outage will last t hours or longer.

Assuming that the available data on service availability has not been mangled and compressed beyond utility by crunching the data to express service availability as a single number, an OCC for service outages can be created empirically from data that defines AV, in exactly the same way that an OCC for service interruptions is created from data defining AC. The durations of outages are sorted in ascending order, and the total operating time observed and total number of outages are used to produce a table like that shown earlier in Table 6.2.

Equivalently, an OCC for service availability can be defined mathematically. To do this, let:

- T_s = the total amount of time that a service is observed for outages;
- T_x = the total time in outages occurring during scheduled and announced maintenance windows; and
- T_o = the total time accumulated in all service outages over that period.

Then the mean time between service outages (MTBO) to be plotted on the y-axis as a function of values on the x-axis representing the outage duration is calculated for any value of x by setting:

$$MTBO[x] = (T_s - (T_x + T_o))/(1 - AV[x]) \qquad (3)$$

Once MTBO[x] is known from service availability data, the corresponding y-axis values to be plotted against x in the OCC for service interruptions is then just:

$$MTBOSI[x] = (MTBO[x])/PP[x] \qquad (4)$$

where the hideous, but pronounceable abbreviation 'MTBOSI', denotes the mean time between operational service interruptions. (The pronunciation, of course, is mitt-bo'-see, with a short "i" and a long "o".)

Since PP[x] is a non-zero probability less than or equal to 1, the effect of division of MTBO[x] by PP[x] in Eq. (4) is a potential increase in MTBOSI[x] to account for the fact that not every service outage will be detected by users as an operational service interruption. PP[x] can be produced as needed from data on the use patterns of the service to estimate *the likelihood that users see a service outage*. The key parameters for doing this are estimates of: *service seizure rate (SSR)*; and the *average duration of connection attempts (CA)*. Their use in estimating PP[x] is seen as follows:

SSR. A seizure of a service is a registered attempt to set up a connection, most commonly executed by going off hook to try to get a dial tone. The service seizure rate (SSR) is defined as the average number of seizures, or log-ons to the service expressed as a rate, usually as the average number of seizures per hour. For most intermittently used services, such as switched voice or data exchanges over the Internet, there are significant and substantial variations in SSRs both with time of day and day of the week. It is therefore useful from the outset to think of SSR as being a set of parameters, $\{SRR[i,j]\}$, indexed by $i = 1,...,7$, representing days of the week, and $j = 1,...,24$, representing hours of the day. The SRR for any index can then be calculated independently by retrieving data from N hourly samples of seizures through the service, as will, for example, be readily available from access trunk group peg counts for switched services, and setting:

SSR$[i,j] =$ (total number of seizures during the jth hour
    of the ith day of the week)$/N$ (5)

In what follows, however, the indexing of SSR will be suppressed unless it is necessary; a reference to "SSR" will be understood to be simply one of values from the set.

CA. Any service seizure then initiates an attempt to establish a connection. Such attempts may be successful, resulting in an answer, or unsuccessful, resulting in a network response indicating that the requested connection could not be set up, e.g. because the destination station was already in use, there was no answer after a number of rings, or some condition made it impossible to complete the connection. The average duration of such connection attempts (CA) is defined as the average amount of time that connection attempts were active, as measured by the time lapsed between seizure of the service for purposes of trying to set up a connection to the disconnection of the seizure by the user, *regardless of the outcome of the attempt.* For most commercial services, ample data for calculating CA will be available from call detail records used for billing. Discrimination by time of day and day of week is generally unnecessary.

Application. An expression for PP$[x]$ denoting the probability that a service interruption will be detected as an operational service interruption as a function of CA and SSR can be developed on the premise that a service interruption will be detected as an operational service interruption only in the event that: (a) there is at least one seizure active at the time of occurrence of the outage, T_0 or (b) there is no seizure active at the time of occurrence, but the next seizure occurs before $T_0 + x$. Under this assumption PP$[x]$ can be estimated by setting:

$$PP[x] = 1 - (P_N)(P[S > x])$$ (6)

where P_N is the probability that there is no seizure active at time T_0, and $P[S > x]$ denotes the probability that the next seizure occurs at a time greater than $T_0 + x$.

The estimate of P_N is derived from CA and SSR by appeal to the relationships illustrated in Figure 6.3. Since SSR is the constant rate of seizures, the average time between consecutive seizures is 1/SSR. Thus, when CA $>$ 1/SSR there is a negligible chance of a gap between the end of one seizure attempt

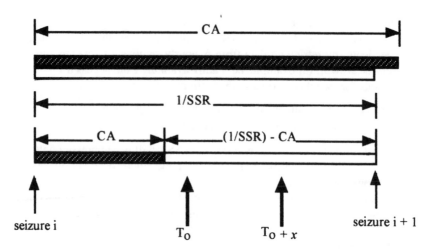

Figure 6.3 Relationships defined by CA and SSR

and the start of another, suggesting that P_N is very close to 0. However, when CA < 1/SSR, there will on average be a gap between the time that one seizure attempt is dropped and the next starts. Since the portion of the time interval of average duration 1/SRR that is covered by an active seizure is CA, this means that:

$$P_N = 1 - (CA)/(1/SRR) = 1 - (CA)(SRR) \tag{7}$$

As is also illustrated in Figure 6.3, when an outage occurs at a time, T_o, that there is no active seizure and lasts until $T_o + x$, it will remain undetected only when the next seizure occurs at a time greater than x. The probability that an outage occurring and correcting before it is detected, given that it was not detected when it first occurred is then derived as follows:

- Since seizures are random events with a constant rate of occurrence, SSR, it is well-known that the probability that the time between the two consecutive seizures shown in Figure 6.3 will *not* exceed time t is

$$1 - \exp[-(t)(SSR)] \tag{8}$$

where exp[x] denotes the exponential function of x, calculated by raising the mathematical constant e to the xth power. The model involved here is illustrated in Figure 6.4, which shows the probability that the time between two consecutive seizures will be more than a particular value for various different values of SSR. Figure 6.4 shows, for example, that when SSR is

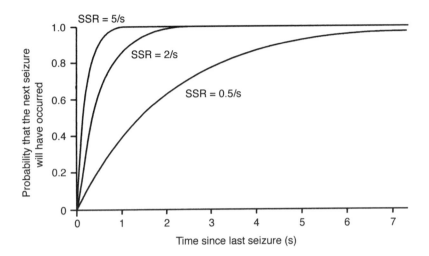

Figure 6.4 Probability Distribution for Time between Seizures

0.5/s there is about an 80% chance that the time between consecutive seizures will be less than 4 s, whereas if the seizures are being generated at the ten times faster rate of 5/s it becomes almost certain that there will be a second seizure within 1.5 s of the last seizure that occurred.

• Since we are interested in the event that the time of the next seizure exceeds the $T_0 + x$, the configuration shown in Figure 6.3 shows that for an outage of duration x occurring when no seizure is active to be corrected before it is detected by the next seizure attempt, the time between ith and $i + 1$st seizure attempts shown in Figure 6.3 must be greater than x+CA. Application of (8) therefore produces the result that:

$$P[S > x] = \exp[-(\text{SSR})(x + \text{CA})] \tag{9}$$

To obtain the desired result, then, it remains to apply (6), (7), and (9) to all the 168 (=7 × 24) possible day of the week and hour of day combinations and average them out to get:

$$\text{PP}[x] =$$

$$(1/168) \sum \sum 1 - \{1 - (\text{CA})(\text{SSR}[i,j])\}\{\exp[-(\text{SSR}[i,j])(x + \text{CA})\} \tag{10}$$

where $\sum \sum$ indicates that the sum is to be taken over all possible combinations of i and j.

Eq. (10) shows, then, that the likelihood that a service outage will be perceived as a an operational service interruption depends both the duration of the outage, x, and on the relationship between CA and SSR. In particular, if CA is greater than 1/SSR[i,j], or very close to it, then any outage regardless of duration will contribute to PP[x]. If CA is small relative to 1/SSR, so that (CA)(SRR) is very small, then the contribution of an outage to PP[x] will depend largely on its duration relative to 1/SSR. Thus, while the calculation shown in (10) is complex, it is fairly easy to test relative values of CA, SSR, and x to determine whether the possible differences for particular combinations of time of day and day of week warrant exact calculation.

6.2.5   Adjustments of OCC Axis Values using Δ[SO,SI\ty]

The other kind of adjustment of the values used to create an OCC for service outages that may be necessary to produce an OCC for service interruptions is an increase or decrease in the recorded durations of the service outages to account for differences between the time an OSI is detected by a service user and a service outage is detected by the service provider. The principal differences that may have to accounted for in this step are:

1. *Reporting delay.* For many services, a service outage may be first detected and reported to the provider by the users. When this happens, it is a common practice for the provider to start the clock on the service outage at the time of receipt of the outage report, and stop the clock when service is restored. When this happens the recorded durations of outages are shorter than the durations of the corresponding service interruptions experienced by the user by the amount of time lapsed between user detection of the outage and receipt of the user outage report by the service provider. The average of these differences is referred to here as the *reporting delay for outages (RDO)*. The RDO for a particular service must be determined directly, by analysis of details of outage reports received by the service provider.

2. *Outage latency.* As suggested throughout the preceding discussions, in intermittently used services there may be substantial differences between the actual time of occurrence of a service outage and detection of that outage as an operational service interruption. The difference is the *outage latency (OL)*, expressed as the expected time between the occurrence of a service outage and its first manifestation to some user as a service interruption. Where necessary to make the adjustment for outage latency for a given service because it is significant, an adequate estimate can be derived from the values of CA and SSR by observing that the latency is non-zero only when there is no seizure attempt active at the time of occurrence of the outage which happens with probability P_N of Eq. (7), and the average

duration of latency when it is non-zero is, as shown in Figure 6.3, just $(1/2)[(1/SSR) - CA]$. This shows that:

$$OL = (1/168) \sum \sum (0.5)\{(1 - (CA)(SSR[i,j])\}\{(1/SSR[i,j]) - CA\} \quad (11)$$

In general, the adjustment appropriate for converting the expected duration of service outages into duration of service interruptions will depend on the nature of the outage and the way the duration was measured. The possibilities for such variations are the reason for including the parameter ty in the expression, $\Delta[SO,SI|ty]$ used here to denote the expected difference between the duration service outages and service interruptions. For example, in view of the possibilities just described, the amount by which the duration of an outage must be adjusted to accurately reflect the expected duration of the corresponding service interruption depends at least on the type of data used to quantify the duration of outages. Assuming that this were the only difference, the two possibilities would be distinguished via the ty parameter by setting:

- $ty = 1$, for outages whose duration is calculated from the time the service provider is first notified by a user that the outage has occurred; and
- $ty = 2$, for outages whose duration is calculated from the actual time of occurrence,

thereby enabling discrimination of the difference between more commonly available outage data derived from user reports and outages assigned $ty = 2$, because, for example, the outage was first detected by the service provider, or data from alarms or monitors were researched to pin-point the exact time of occurrence.

Use of this notational convention to define the appropriate values produces: $\Delta[SO,SI|1] = RDO$, and $\Delta[SO,SI|2] = -OL$. The distinction of the measurements by use of the ty parameter also enables the calculation of the best estimate from a mixed bag of measurements of outage durations by setting:

$$\Delta[SO, SI] = \sum (Pr[ty = i])(\Delta[SO, SI|i]) \quad (12)$$

where $Pr[\cdot]$ denotes the proportion of items in the sample with characteristic $\cdot$, and Σ is understood to indicate summation over all possible values of the index i.

Once the final expected value of $\Delta[SO,SI]$ has been determined, that difference is incorporated into the construction of the OCC for service interruptions from the data used to construct an OCC for service outages, by setting:

$$MTBOSI[x + (\Delta[SO, SI])] = (MTBO[x])/PP[x] \quad (13)$$

This means, for example, that if $PP[x]$ were 1, and $\Delta[SO,SI] = \Delta[SO,SI|1] =$

RDO, because all outage durations were calculated from the time each was reported to the service provider, then an outage of duration x would have actually appeared to the users as an operational service interruption of duration x + RDO, indicating that the corresponding value of MTBO would actually apply to a much longer MTBOSI value.

6.2.5.1 Implications

Any readers who read books like this like I do will have skimmed over the preceding material, making mental notes that there were some complicated looking definitions and equations that describe a process for converting an OCC for service availability into an OCC for service accessibility, by doing some failure rate somethings with other latency things to get MTBO something or the other. This is fine, because all of these details were put here not so much for education, but to demonstrate, by actual accomplishment two very important points:

1. It is possible to generate an OCC for service interruptions from an OCC for service outages; and
2. Such a transform for a particular customer cannot be accomplished without consideration of that customer's patterns of usage of the service.

These characteristics are very important, because they demonstrate the potential for building a very important bridge between the perspectives of service providers and service users. An OCC for outages displays what a service provider knows. What users of an intermittently used service really want to know, however, is something that is specific both to how the service is used and how disruptions may affect that use. The service availability OCC, or worse yet an expression of the average service availability, responds to that need by saying, in effect: "Here's what we know. You figure out for yourself whether that represents acceptable quality". An OCC for service interruptions, tailored by using data on a particular customer's use patterns to transform the availability OCC conveys an entirely different message: "We know that you are concerned with the frequency and duration of service interruptions you will experience, rather than our performance. Accordingly, we have taken the extra step of analyzing the way you will use our services to give you a way of understanding what to expect and gauging for yourself how often what is experienced will represent unacceptable service".

I am sure that I, as a customer, would have no difficulty deciding which message that I would rather receive, given the choice. However, I might at the same time be somewhat chary of the results, especially if it looked as if the MTBOSI OCC might be substantially discounting some of the effects shown in representations of service availability. It is therefore important to be able to appeal to the kind of models detailed earlier to get something that looks sensible and shows that not only do you understand the customer's particular concerns better than the competition, but you have been careful to get it right...

It should also be noted in this context that the ability to transform OCCs for service outages to OCCs for operational service interruptions suggests that it will be doubly useful to make OCCs the quantifier of choice for analysis of intrinsic QoS with respect to accessibility. Adoption of the OCC as the quantifier for availability, for example, will foster perception of the importance of maintaining data on both the frequency and duration of outages, while other quantifiers might encourage the fatal mistake of calculating and archiving availability ratios.

6.2.6 Evaluation

The point of using OCCs for operational service interruptions to quantify service accessibility is that they characterize performance in a way that gives the service users the ability to execute the final step in analysis of service accessibility for themselves, by evaluating that performance in light of their perceived needs to determine whether the QoS is acceptable. This quantifier for accessibility therefore achieves the highest art in decision support by providing to the decision-maker precisely the information needed, instead of a conclusion derived from someone else's interpretation of the available data.

In addition, OCCs are in general very useful interpretation aids that can be readily used to support identification and assessment of differences in performance with respect to occurrence of any events of different durations in such activities as:

- Comparing the quality of two different services;
- Detecting the occurrence of significant changes in quality and identifying likely causes of those differences; and
- Routinely monitoring and evaluating intrinsic QoSs.

6.3 Continuously Used Services

Although intermittently used telecommunications services like switched voice and Internet access are the most visible and familiar ones because their use is interactive, there are also telecommunications services that are less visible, but are expected to be continuously available for use, day in and day out, 24 h a day, such as e-mail and bulk data transfer services. Like intermittently used services, such continuously used services may employ dedicated facilities, such as a dedicated computer-to-computer connection, or switched facilities, such as those used for store-and-forward message relay, so implementation is not a discriminator. Rather, their distinguishing characteristic is that there is a nearly constant demand for use of the service, so that any unscheduled interruption of any connection is immediately felt, unexpected, and therefore automatically an operational interruption.

6.3.1 Concerns

Because of the differences in use, the concerns with accessibility expressed by users of intermittently used services are of much less interest to users of continuous service than the effects of operational interruptions on the ability to keep pace with the continuously presented demand. The basic question of accessibility to be addressed for continuously used services is consequently something like:

> What is the *relative* effect of connection failures on the information transfer capacity of the service?

The word 'relative' is highlighted here, because interruptions in service are in this case but one of many independent factors that determine the effectiveness of a continuously used service. It therefore makes no sense to try to gauge the effects of connection failures as if they were the only contributing factor.

6.3.2 Measure

The recommended measure for evaluating perceived QoS with respect to accessibility for continuously used services is the *operational effective capacity (OEC)* of the service, defined generically by the ratio:

$$OEC = DEX(t)/CAP(t) \qquad (14)$$

where $DEX(t)$ is the amount of user injected information exchanged over the service during some span of time lasting t time units, and $CAP(t)$ is the maximum rated capacity of the service over a span of time lasting t units, expressed in the same units as DEX.

Suppose, for example, that the service being analyzed is a continuously used ISDN connection rated at 128 kbps, and we observe the use of that service over a period of $t = 24$ h, finding that 7.95 Gbits of injected information are exchanged over the service under a continuous load. Then the rated capacity of the service for that period would be 11.059 Gbits (24 h × 3600 s × 128 kbps) of data, so the OEC would be 71.89% (= 100 × 7.95/11.059).

Suppose, further, that during this time period there was an outage lasting 1 h, so that the connection availability was 95.83% (= 100 × 23/24). Then we would find that during the time the connection was up the OEC was still only about 75% (= 100 × 0.7189/0.9583), indicating that about 25% of the available capacity is being devoted to encoding overhead, or being used for retransmission of data blocks that are not received error free the first time they are transmitted. This means, in turn, that the effect of concern in measuring accessibility for this hypothetical service is most accurately characterized by indicating that 1 s of outage of the service results in a reduction in the capacity for transfer of user injected information by 96, rather than 128, kbps.

6.3.3 *Quantifiers*

As suggested by the generic definition, the OEC for any continuously used telecommunications service can be estimated directly by:

1. Determining from design or engineering of the service the value: ι = the rate at which the service transmits information units; and
2. Observing operations over some period of time, recording: T_s = total amount of time the service was observed; T_x = total amount of time the service was interrupted for scheduled maintenance or reconfiguration activities; and U_i = total number of user injected information units transmitted via the service during the period of observation.

The amount of time the service was expected by its users to be up and usable is then $T_u = T_s - T_x$, and when the rate ι is expressed in the same units of time as T_x, T_s, and T_u, a direct estimate of the OEC is given by:

$$\text{OEC} = (U_i)/(\iota)(T_u) \tag{15}$$

The efficacy of this estimate depends on the validity of the assumption that the service was continually loaded by a demand at least as great as the capacity during the period, T_s, that its operation was observed, and verification of this condition may sometimes be difficult. It is therefore usually much easier in practice to estimate OEC indirectly, by setting:

$$\text{OEC} = [(CA)(TE)]/[(1 + HO)(1 + EO)] \tag{16}$$

where CA is the connection availability, TE the service throughput efficiency, HO the handling overhead, and EO the encoding overhead, defined and illustrated as follows.

6.3.3.1 Connection Availability (CA)

For intermittently used services, the appropriate measure of availability was *service availability,* for which a failure represents a condition under which none of the connections normally supported through the service can be established, and the single value quantifier is the expected proportion of time that *all* attempts to establish connections normally supported through service will be unsuccessful. For continuously used services the appropriate measure of availability is *connection availability,* which reflects the expected availability of any one of the connections normally supported by the service. It can be quantified by selecting n of those connections, indexed by i, observing each for a period time, $T_s(i)$, to record:

$T_a(i)$ = the total time the ith connection was in use, or established and

available for use, but idle due to lack of demand, and

estimating CA by the ratio:

$$CA = \left(\sum_{i=1}^{n} T_a(i)\right) \bigg/ \left(\sum_{i=1}^{n} T_s(i)\right) \tag{17}$$

With such an estimate in hand, the problem of verifying the constant loading on the service in order to user the direct estimate in Eq. (16) is circumvented by using Eq. (17) as the estimate that a connection will be usable, and characterizing the other performance factors with quantifiers based only on the traffic carried.

6.3.3.2 Throughput Efficiency (TE)

The first performance characteristic to be sampled in this way for use in conjunction with CA is the *throughput efficiency* for the service, defined generically as the ratio of the throughput of data actually achieved over an usable connection to that which would have been achieved had there been no errors in transmission. The bases for quantifying throughput efficiency for any service are *transmission units,* representing the smallest block of data that can be requested for retransmission by the destination in order to correct errors received in the original transmission, and *information exchange units,* representing complete messages or clumps of data to be delivered to a particular destination. For e-mail services, for example, the information exchange unit is an e-mail message and its attachments, prepared by the user for delivery as a unit to particular destination(s); the transmission units are blocks of characters of some relatively small fixed size. In bulk data transfers, the information exchange units are data files, which are transmitted in units comprising data blocks of a small fixed number of bits.

In either case, each transmission unit is checked for errors at the destination by application of character or error detection and correction coding. The destination then either sends back acknowledgment of receipt for blocks in which no errors were detected, or requests re-transmission of any blocks that were found by the far end detection routines to have errors. The blocks that are retransmitted under these procedures then contribute to the denominator of the ratio defining throughput efficiency without increasing the numerator, thereby reducing the value of the ratio.

In addition, since far end error detection routines may not be implemented in a service, and those that are cannot possibly detect all errors incurred in transmission, there is a secondary effect of errors in transmission, in that the undetected errors in an information exchange unit may render the information contained unreadable or unusable because of ambiguities in, or loss of, critical elements of information in the information exchange unit. Errors in transmission may, therefore, force the intended recipient of an information exchange unit to request retransmission of the whole thing, even more seriously degrading the throughput efficiency.

Since the throughput efficiency depends only on the traffic carried, it can be quantified directly by observing transmissions for some fixed time period and recording:

- M = number of information exchange units transmitted;
- U_i = the number of transmission units in the ith information exchange unit; and
- U_T = the total number of transmission units transmitted during the period observed.

Then the throughput efficiency, TE, can be estimated by setting:

$$\text{TE} = \left(\sum_{i=1}^{M} U_i \right) / U_T \qquad (18)$$

Alternatively, if reliable, large sample estimates of:

- R_{tu} = probability that a transmission unit will be retransmitted, and
- R_{iu} = probability that an information exchange unit will have to be retransmitted

can be produced (as is usually the case in the data-rich environment of telecommunications), then throughput efficiency can be quantified indirectly by the ratio:

$$TE = 1/[(1 + R_{tu})(1 + R_{iu})] \tag{19}$$

6.3.3.3 Handling Overhead (HO)

When continuously used services involve exchanges of information without the active oversight of the originator, instructions for routing, handling, and delivery must accompany the information to be delivered to the destination(s) all the way through the system. The bits or characters added to the information exchange units in order to accomplish this are exemplified by the message headers that are seen on recipient copies of e-mail messages and the sometimes very long identifiers that are appended to data files exchanged via bulk data transfer services that are used and stripped off by the file server. The resultant increase in the size of information exchange units is called here the *handling overhead* for the service, and is defined generically as the expected proportional increase in the size of information exchange units mandated by the exchange unit formatting requirements of the service.

Because it is determined by the number of destinations for information in each information exchange unit, and the numbers of characters or bits required under the formatting conventions for a particular service, handling overhead is highly variable. It can, however, be readily quantified for a particular service by gathering a large sample of information exchange units, indexed by i, and determining for each:

- H_i = the number of transmission units, bits, or characters used to specify handling in the ith exchange unit sampled; and
- T_i = the corresponding total number of transmission units, bits, or characters in the ith information exchange unit.

Then, in terms of these values, the expected handling overhead can be expressed as:

$$HO = \left(\sum H_i\right) \Big/ \left[\left(\sum T_i\right) - \left(\sum H_i\right)\right] \tag{20}$$

which is hardly worth mentioning, except as a means of underscoring the point that handling overhead is a measure of the effects of requirements for formatting of information exchange units on overall throughput, rather than on individual information exchange units (or so said the Cheshire cat as it dissolved into a smile).

6.3.3.4 Encoding Overhead (EO)

Handling overhead for a service, then, reduces the efficiency of information exchange by increasing the size of information exchange units. Encoding overhead for the transmission protocol for a service does the same thing for transmission units. The overhead in this case is defined to be any additional bits or characters mandated by the transmission protocol for purposes of: identifying the information exchange unit associated with each transmission unit; framing, sequencing, and controlling the handling of the transmission units; or supporting forward error detection and correction.

The encoding overhead is defined generically as the proportion of each transmission unit that carries encoding, rather than contents of an information exchange unit. The proportion in this case is a constant, which can be derived directly from specifications of the transmission protocol.

6.3.4 Evaluation

The implication here for evaluation of the accessibility of continuously used services is, then, this:

> It is impossible to meaningfully address concerns with accessibility of continuously used services without considering all factors affecting the information exchange capacity of the service.

To see this, consider the following example, drawn from a true-life experience. Only the names have been changed to save the guilty from embarrassment.

One day, back the early 1980s I was talking off-the-cuff to one of the larger Satellite Business System customers, who was using our satellite transport for a continuous use data exchange system. The operations manager of this service happened to mention that she had been approached by a competing telecommunications service provider, BS&S, who had presented her with their data showing that the operational availability of their terrestrial data service was 99.99%, while our satellite service was operating at about 99.95%, meaning that we were down on each link about 3.5 h more per year than they were.

She allowed that the difference was not enough to be an issue, but she wondered if I could give her any help, because the marketers for the competing service were using that difference to create doubts as to the efficacy of the choice of this new satellite service in the minds of her superiors. My advice was for her to go back to the sales people for BS&S and ask them to produce figures on the throughput efficiency for their services to be compared with the figures that she could readily derive from the traffic flow monitors on their data links through the satellite.

They came back with a throughput efficiency of about 97.5% (remember this was back in the early 1980s, before there was much optical fiber transport), while she had calculated a throughput efficiency better than 99.99% for the satellite links (which were maintaining bit error rates of 10^{-9}–10^{-11} when they were up). Thus, she didn't even have to multiply to see that the errors on the terrestrial transport were impeding flow of data in a way that represented a loss of throughput equivalent to the results of outages of about 219 h a year on each link.

I could go on and on endlessly citing examples of such possible trade-offs among connection availability, throughput efficiency, and handling and encoding overheads, but the point should by now be clear, and overstated, by the advice that evaluations of continuously used services should be based on comparisons of the operational effective capacity any time there is a possibility of a difference in performance with respect to more than one of these factors...

7

Routing Speed

7.1 Evaluative Concepts

As described in Part I, the establishment and maintenance of any connection through a switched telecommunications service requires: (1) dynamic determination of precisely what node-to-node links will be used for a requested connection; and (2) seizure and interconnection of the selected node-to-node links. The speed with which this process is effected to set up a requested origin-to-destination connection is referred to here as the *routing speed* for a service.

The effects of such routing speed may be manifested to users in one of two ways, depending on whether the service employs circuit- or packet-switching. The essential differences in effects are seen as follows:

- *Circuit-switching*. Circuit switching is a technique whereby connections are set up on request by selecting node-to-node links that are reserved exclusively for exchanges of information between the specified origin and destination. Once established, the circuit switched connection is supposed to stay up until there is a signal from the users for it to be taken down. The most familiar circuit-switched services are those that are intermittently used, such as dialed-number telephony. In using such services, the users must actively monitor each attempt to set up a connection, and be prepared to initiate use when the connection is established. Such active oversight of attempts to effect the node-to-node routing from origin to destination exposes users to, and creates an awareness of, a visible manifestation of the routing speed of the system.
- *Packet-switching*. Packet switching is a version of store-and-forward switching in which the information to be exchanged between origin and

destination is subdivided into very short segments, and the node-to-node links from the origin to destination are selected dynamically and independently for each such "packet" conveyed. Although this kind of switching does not require active monitoring of the selection process by the user, the process may affect both the delay and the variance in the delay in origin-to-destination transmissions in ways that become noticeable to users.

In what follows, either type of switching is examined in turn to describe the appropriate methods for measurement of routing speed and evaluation of its effects on user perception of QoS.

7.2   Circuit-Switched Services

7.2.1 Concerns

Although there are other services of this type, the most readily recognized circuit switched services are the familiar dial-up telephone services. Accordingly, in what follows, the terms "dial" and "dialing" will be used to describe the process of requesting establishment of a particular connection, even though that request might entail something else, such as a "hailing" message.

The process by which a connection is established the through dial-up telephone services we are using for the model of circuit-switched services, then, is this: the user picks up the telephone hand-set to signal the desire for a connection, and dials the destination station number in the proper format to specify the particular connection desired. Upon completion of dialing, the user begins to listen for audible responses from the telephone service that will indicate the disposition of the attempt to set up a connection to the destination. The person placing the call is at this point particularly interested in ring-back signals, which will indicate that the connection has been extended to the destination station, suggesting that the user should prepare to initiate exchanges of information when the called party or device answers.

All users of circuit-switched services, even those who have assistants who routinely place calls for them, thus spend some time listening after a number is dialed for indications of what is happening with respect to the request to set up a connection, and will, over time, synthesize this experience to develop expectations as to how long it will take before various possible responses are received. In particular, that synthesized experience will become the basis of the users' deciding when something has gone wrong with the call attempt, because there has been an inordinately long wait without any response whatsoever.

The synthesis of experience in using the telephone and monitoring the line after dialing, in turn, creates a consciousness of the time lapsed after dialing that fosters two basic concerns with respect to routing speed of a circuit-switched service:

How long does it take before I know that a requested connection has been extended to the destination?

Is this time stable and predictable?

7.2.2 Measure

The most general measure of routing speed for circuit-switched services is the *post-request delay*, defined generically as the elapsed time between the time that a user completes the request for connection, and the time of receipt of the first response back indicating the disposition of the connection request. In the familiar case of dial-up telephony, for example, this time is commonly referred to as the *post-dial delay (PDD)*. It is defined generically as the time elapsed between dialing the last digit (or symbol) of a telephone number and the receipt of the first audible network response indicating whether the requested connection will be completed.

The types of audible signals that might be distinguished by the user in the perception PDD include:

- SBY – *slow (station) busy*, usually pulsed at 60 ips (impulses per second). This signal indicates that the distant station called is off hook or "busied out" by the local service provider. Users will presume that this signal means that the distant station is in use. However, as was described earlier, a station busy signal will also be generated when the station set is inadvertently off hook or has been temporarily disconnected by the local service provider to compensate for an inadvertent off hook condition.
- RDR – *reorder (network busy) signal*, usually transmitted at 120 ips. This signal is supposed to indicate that the requested connection cannot be made because there is no available facility for effecting one of the node-to-node links needed to complete the connection. It may, however, also be transmitted when the switching system determines that the number dialed cannot be routed, as happens, for example, when the proper number of digits needed to route a call is not received, or is not received quickly enough by a switching device, or the set of digits received do not correspond to any known station. The reorder signal will frequently be followed by an recorded voice announcement explaining the problem.
- SIT – *special information tone*, comprising a three-tone warble. This tone usually indicates some problem with the number dialed making it impos-

sible to route, and is frequently followed by a recorded voice announcement indicating that the call could not be completed as dialed.

- *RVA – recorded voice announcement.* Sometimes the first response received will be a recorded voice announcement, without the preceding RDR or SIT tone, explaining a condition that is preventing completion of the requested connection. (For example, "Due to the unusually high traffic volume on Mother's Day we cannot complete your call at this time. Please try again later".)
- *RNG – ring-back signal*, indicating that the connection has been set up to the destination station, and the user had better get ready to talk. Ring-back signals usually, but not necessarily, have a cadence of 4 s of silence followed by 2 s of signaling, comprising one or two pulses of a tone.
- *ANS – station answer.* Because of the high-speed routing achieved with modern out-of-band signaling it is also possible that the first indication of completion of the connection will be an answer by the party or device called.

In addition to these audible responses from the network indicating what has happened with respect to the routing of the connection request, it may happen that the system loses track of the request, and nothing at all is signaled to the user. When this occurs, the connection attempt is described in the vernacular of telephony as having gone *high-and-dry*. The occurrence of this kind of failure is usually signaled to the user by one of the completion failure signals listed above after a very long delay.

7.2.3 Quantifiers

Quantification of the PDD experienced by users of circuit-switched, dial-up telephony is an exercise in timing of call progress for a sample of call attempts. Although the delays incurred when other types of audible signals are heard might be timed and analyzed, the convention for direct quantification of PDDs is to consider only those call attempts for which the first audible response was one indicating that the desired connection was set up. This is accomplished by sampling call attempts to obtain a set of attempts, indexed by i, for which RNG or ANS was the first audible response, and recording for each:

$$\text{PPD}(i) = T_a(i) - T_d(i) \tag{21}$$

where $T_d(i)$ is the time that the last digit was dialed on the ith call attempt and $T_a(i)$ is the time of first detection of the audible response to the ith call attempt.

The most useful quantifier for characterizing PDDs is the *frequency distri-*

bution of all PDD values observed in a sample of delays as defined in Eq. (21). Unlike the meaningless average of these measurements of PDD, the frequency distribution of the measured delays for a large sample will generally look something like Figure 7.1, exhibiting delays that cluster about multiple *modes* (i.e. high points) in the distribution, each of which reflects the routing speed associated with one of possibly many different types of routes that might be used to set up the desired connection.

The reason for using the frequency distribution as the principal quantifier for PDD is similar to that for using the OCCs rather than simple ratios as the quantifier for accessibility – there are two questions to be answered in addressing any user concern with PDDs. The first question is "what magnitude of delays can I expect?", which might be answered by a simple description of the distribution. The second, however, is "how predictable are they?" which can be meaningfully addressed only by appeal to a graph like that in shown in Figure 7.1. This graph was constructed from the data shown in Table 7.1, which is purely hypothetical, but typical of the distribution of PDDs. It clearly shows that the PDDs for this hypothetical route are tightly clustered about 2, 4, and 6 s, and are therefore much more predictable than suggested by a description of the PDDs as having average value of 2.9 s, a standard deviation of 1.4, and a range of about 1.2–7.0 s. Rather, the most useful short description of the frequency distribution shown in Figure 7.1 (if there is one) is that: PDDs over the hypothetical route exhibit three modes, at 2, 4, and 6 s. The distributions about these modes comprise: a cluster with an average of 2 s and standard deviation of 0.3 realized for 65% of the connections; a cluster with an average of 4 s and standard deviation of 0.2 realized for 25% of the connections; and a cluster with an average of 6 s and standard deviation of 0.3 realized for 10% of the connections. (As I said, *"if* there is one…"*).

Table 7.1 Sample of measurements of PDD (s)

1.9	2.0	1.9	4.0	1.6	5.9	2.1	4.0	4.1	2.3
2.2	1.7	4.1	1.7	2.2	1.8	2.0	2.0	2.1	1.9
2.2	1.8	2.0	4.0	2.0	7.1	3.4	1.9	3.1	5.8
2.2	2.0	2.1	4.0	6.6	1.6	3.9	4.0	2.3	2.1
2.0	4.1	3.9	2.2	4.0	2.3	2.7	4.0	2.0	1.9
2.2	4.0	2.0	1.8	3.8	2.1	2.1	1.9	5.4	1.0
4.0	1.9	2.2	1.9	1.9	2.5	1.7	4.0	1.9	4.0
6.0	2.4	1.2	4.0	2.2	1.5	1.8	2.1	5.9	1.7
6.1	2.0	2.0	4.3	6.2	4.0	2.4	1.8	2.3	1.8
3.9	6.0	2.4	4.0	4.1	2.0	2.1	2.1	4.0	2.0

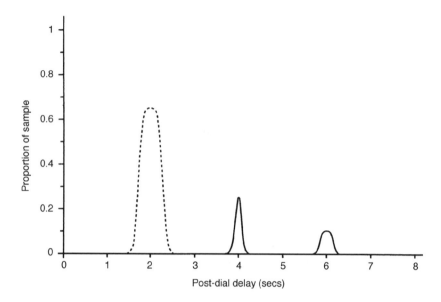

Figure 7.1   Frequency distribution of PPDs for a hypothetical origin/destination pair

The direct measurement of PDDs in the manner just described requires a cooperating station to which a number of calls can be placed. This can usually be arranged when there are relatively few call attempts involved. However, there may be cases in which the PDD to a particular destination is to be measured from a number of different origins, and the testing will begin to interfere with the normal use of the service. When this happens, a convenient alternative to measurements to a distant station will be measurement of routing speeds in calls into a responder in the switch terminating the destination. The measurements in this case will be conducted by placing calls to the responder to obtain a set of attempts, indexed by i, for which the first audible response was the responder *test progress tone* (TPT), and recording for each:

$$T_r(i) - T_d(i) \tag{22}$$

where $T_d(i)$ is the time that the last digit was dialed on the ith call attempt, and $T_r(i)$ is the time of first detection of the TPT.

The differences in Eq. (22) are sometimes referred to as *call set up times* (*CSTs*) to distinguish them from the PDDs that are perceived by users.

It is important to recognize that the delay times defined by Eq. (22) differ from those experienced by users in that the TPT is returned as soon as the responder line is seized, circumventing the delay in setting up a ringer in a line and generating the first audible ring-back signal. The additional delay is referred to as the *ring-back latency (RBL)*. It has two components:

- *Ringer connect time (RCT)*: the time it takes to connect a ring signal generator to the origin/destination circuit after the line to the destination station has been seized; and
- *Ring signal latency (RSL)*: the expected amount of time before ring signal energy will be heard by the originator after a ringer has been attached.

The ringer connect time is determined by the design and operation of the circuit switch terminating the called station. It is relatively short, usually somewhere in the range of 0.1–0.5 s, and stable enough to be to all intents and purposes a constant.

The ring signal latency depends on the type of ring-back signal generator employed at the switch. The two types commonly employed are illustrated in Figure 7.2. The first unit shown there is a single signal generator, which is attached to every line required to transmit ring back. This unit provides a 2 s surge of signal power followed by 4 s of silence. When it is attached to a line there is an instant ring-back signal with probability 1/3 (= 2 s of power/6 s total cycle), and no power with probability 2/3. Given that there is no power, the average wait until the next ring-back signal starts is 2 s. The average RSL for the single ringer configuration is therefore 1.33 (= (2/3) × 2) s.

The other configuration shown in Figure 7.2 is a bank of three ring-back signal generators, timed so that each is generating 2 s of signal power in a different part of a 6 s cycle. When a ring-back signal is required, the system checks the status with respect to the 6 s cycle, and attaches the signal generator that is next in line to generate signal power. The RSL in this case is 1 s.

7.2.4 Evaluation

Because it is readily estimated and a characteristic that is visible to users, the PDD for a manually dialed circuit-switched service is frequently one of the first performance characteristics specified in statements of requirements prepared by representatives of large user communities. The "required" PDD cited is sometimes expressed as if PDD were some hard-edged, immutable service attribute, and failure to achieve a particular PDD will result in widespread rioting in the user community. However, as illustrated by the dialog in the box below, nothing could be a more poorly conceived expression of what is important in evaluating PDD of a particular service.

User: So tell me, how fast will my calls go through after I've converted to your service?

Analyst; Well, that depends...

User: Depends? Depend on what? Your network is using this modern out-of-band signaling. Doesn't that mean that all calls will have about the same set up time? Your competitor tells me that the call set up time in his network is x seconds on average. How does yours compare with that?

Analyst: Oh. we're the same, plus or minus a tenth of a second or so.

User: So, that x seconds is the kind of time I can expect for my calls. Right?

Analyst: Well, yes, no, and maybe. The first thing that you have to realize is that the call set up time is not what you're going to perceive. At the least there will be an additional delay while the distant station attaches a ringer to the line. This may take as much as half a second. and after that there may be an additional delay until you hear the first ring on your line.

User: So that adds...what?

Analyst: Well, that depends. One type of ring system uses three ring generators and attaches any incoming call to the ringer next scheduled; to generate a ring signal. This configuration adds 1 s to the average PDD. A single ringer system adds 1.33 s to the average. But either way there will be cases where what you experience is upwards of 2 s longer than x.

User: OK, OK I get the picture. So what you're telling me is that calls from my sites will go through about as fast as the average PPD you tell me you've calculated for your network in end-to-end tests give or take a few seconds.

Analyst: Well, that depends. If we can convince ourselves that the routes we tested in those tests are pretty much the same and in the same proportions as routes you will be using, then that may be true. However, there are a few other factors that have to be taken into account and verified before I could assert that with any confidence.

User: Oh, come on. What else can there be?

Analyst: Well, it depends. Try these: are any of your sites in rural or sparsely populated areas? If so, then they are likely to be originating and terminating calls through switches that connect to the rest of the network via in-band signaling. That's going to add about 1.5 s to the call set up time for each in-band hop. Are your calls going through PBXs? Then there will be a very big difference in PPD if your PBX holds the digits dialed to determine the least cost routing instead of sending the digits to the next switch as they are received. Are you going to make seven-digit on-net calls? Then there will be an additional delay of about half a second while the seven-digit number is being translated. Do you want a good estimate of what callers into your sites on freephone numbers are going to experience? Then you may have to add another half second to the PDD you are seeing on seven-digit outbound calls, because the 800-numbers are going through two translations, one by the local service carrier to determine which long distance carrier gets the number, and one by the long distance carrier to translate the 800- number into the standard numbering plan. While we are at it, how about your users' dialing practices? If the calls going out of your sites are going through interfaces configured for overlap outpulsing to the next switch, the call set up time itself will depend on the speed and cadence with which the number is dialed. Up to a point slower dialing will result in faster call set up. Oh, and do your users know to add the "#" to manually signal end of dialing, when the number dialed may be seven or ten digits. You're looking at a good 3–5 s swing in PDD on this one!

User: So the bottom line is…?

Analyst: The only thing that average PDD in the tests I showed you is good for is determining whether there are gross differences in routing speeds between two different services. In a network tested as having average PDD of y seconds there are going to be users who are routinely experiencing PDDs averaging 6–9 s longer on their calls, and every user will occasionally encounter PDDs 5–7 s longer than the quoted average value. Moreover, if you are trying to find out how long it will take for international calls to go through all bets are off. Because of factors over which long distance carriers have absolutely no control unless your calls stay on somebody's global network, what you're going to see depends on where you are, what country you are calling into, and where in that country your call is going. Not only that, there will be large variations in both the average and individual PDDs, as your calls are variously routed among the myriad direct and transit routes built, bartered, leased, owned and borrowed to complete telephone calls around the world. One call goes out over a premium route and it's connected before you've even had a chance to stir your coffee and settle in for a wait; the next one to the same location just a few minutes later may go onto overflow routing and wend its way to the destination via so many slow hand-offs that your coffee is beginning to get cold before it connects. So what do I tell you?

Stir your coffee before the call, or pour it hot and fix it after you've
dialed the number? Either way you are eventually going to be disap-
pointed with my advice.

As suggested by the preceding analyst/user exchange, then, there are at least
two good reasons for asserting that average PDD is a poor quantifier for
routing speed:

SINGLE RINGER   THREE BANK RINGER

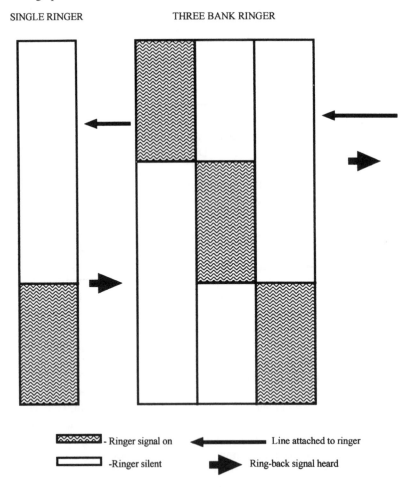

Figure 7.2   Comparison of Responses of Single- and Three-Bank Telephone Ringers

(1) *PDD experienced at an origin will vary with the destination called.* The PDD that can be reasonably expected in any call from an origin depends on how the routing of calls is handled in each of three distinct phases of the set up of the node-to-node connection from the origin to a particular destination. The phases of routing are:

- *Access*, during which links between the origin and the first switch in the long-distance network are selected;
- *Transport*, during which the route across the long distance network to the switch terminating the destination is determined and set up; and
- *Termination*, during which the routing from the last long distance switch to and through the terminating switch (usually referred to as the *end office*) is completed. This phase includes a test of the status of the line(s) to the destination station or PBX to verify that there is a line that can be used, seizure of an available line, attachment of a ringer to the line, and transmission of the first audible signal back to the origin.

Even discounting other possible reasons for variations in PDD, there are manifold possibilities for the way that terminations are effected, depending on the destination called. As a result, the question of what PDDs might be expected in calls originated from an origin cannot be reasonably answered without reference to the specific destination for the call.

Moreover, even the average PDD in calls from an origin cannot be expressed in any meaningful fashion without first determining the mix of different types of routes represented in the collection of destinations called, together with the frequency with which each destination is called, so that the average can be appropriately weighted. Recognition of this sensitivity of average PDDs to the patterns of calling from a particular origin is particularly germane when the calls originated include overseas destinations. There are such wide country-to-country variations in access, transport and termination of calls that a PDD value has no meaning whatsoever without at least specifying the origin and destination countries.

(2) *There may be significant variations in PDD even among calls from an origin placed to the same destination.* Circuit-switched networks are designed to accommodate nearly all requests for origin/destination connections by providing for numerous alternatives for establishing a set of node-to-node links that can be interconnected to effect the connection. As a consequence, even when the delays in access and terminations phases of routing are fixed by the configuration of the service, there are inescapable variations in the routing of a call during transport, depending on which of possibly millions of alternative routes is selected to get to the termination switch.

(3) *Moreover, the emphasis on magnitude of PDD as a requirement for*

quality of service is not well-founded, because users tend to adapt to, and accommodate, PDDs. Anyone who has used the telephone over the last 20 years can readily verify this. With the deployment of out-of-band signaling systems to replace in-band signaling over the last two decades, the nominal PDD for most domestic long distance and many international routes has substantially decreased, dropping in some instances from 10 to 30 s to less than 5 s. Yet, hardly anyone today can remember being perturbed by consistently having to wait an inordinately long time for calls to complete. The reason, I claim, is that users tend to set their expectations of PDDs by what they are currently experiencing, rather than insist that their experience match some pre-conceived expectation. The attitude is pretty much that: "PDD is what it is, so rather than fight it, I'll accommodate it. When things were slow, I would dial the number and then catch a sip of coffee while I mentally rehearsed what I needed to say. Now that there're faster, I grab the sip of coffee, think about what I want to say, and then dial". (If this casual observation is not convincing, try the experiment suggested in the memos in the boxes on the following pages.)

Memorandum
To: All skeptics who doubt that users of circuit-switched services accommodate PDDs
From: A bare-foot empiricist
Subject: An experiment

The hypothesis is that users tend to set their expectations of PDDs by what they are currently experiencing, rather than insisting that their experience match some pre-conceived expectation. To test this hypothesis for yourself, conduct the following experiment:

1. Guess the PDD for your own domestic long distance service. Place some test calls, noting the time that it takes after the last digit was dialed before you hear ring back. Compare your guess with the range of times determined in your test.
2. Armed with the hard data from your informal test, ask any five persons not in the telephone business how long it takes to connect a long distance call, making note of their unprompted answers.
3. For any subjects who initially protest any capability to answer the question, press for a guess, and record the results.

Evaluate the results of your experiment according to the guidelines in the follow-up memorandum.

These modest observations then, suggest two bases for evaluation of QoS with respect to PDD, depending on whether we are interested in perceived or intrinsic QoS.

7.2.4.1 Perceived QoS

In the case of perceived QoS with respect to PDDs in circuit-switched services, the users' assessments will be focussed on PDDs that are *noticeably different* than those *normally experienced*. Both of the italicized conditions here are highly subjective and very "fuzzy", depending on user sensitivity to time differences, and the characteristics of the PDDs of the service to which they have become accustomed. However, for purposes of evaluation PDDs using frequency distributions like those shown earlier in Figure 7.1, three crude rules of thumb can be applied:

1. The range of PDDs that are "normally experienced" is estimated by examining the modes of the frequency distribution, beginning with the mode whose associated values represent the largest proportion of the sample, and moving outward in both directions, adding the proportions of the sample associated with the modes, until either:

   - The total proportion represented exceeds 90%, or
   - The total proportion represented exceeds 75%, and inclusion of the next mode's associated points would cross a gap of more than 5 s. Then, if μ_1 and σ_1 denote the mean and standard deviation of the values about the smallest mode identified in this process, and μ_2 and σ_2 denote the mean and standard deviation about the largest, the range of normally experienced PDDs is $\mu_1 - 3\sigma_1$ to $\mu_2 + 3\sigma_2$.

2. To be noticeably longer, a particular PDD must be 5 s or more longer than, but no more than about 15 s longer than, the longest PDD normally experienced.
3. To be noticeably shorter, a particular PDD must be 3 s or more shorter than the shortest PDD normally experienced.

Thus, for example, to apply these rules to the frequency distribution of PDDs displayed in Figure 7.1, we would start with the mode at 2 s, to pick up 65% of the distribution, then move up to the mode at 4 s, to pick up another 25% of the distribution, making up the requisite 90% of the sample. On the basis of this determination, the imputed range of normally experienced PDDs would 1.1–4.9 s. Given this range for normally experienced PDDs, a particular PDD would have to be about 9.9 s or longer to be classified as noticeably different. This means that the 10% of the PDDs clustered about 6 s are different from those imputed to be normally experienced, but not different enough to be noticeable to users.

Similarly, anything above 19.9 s would be excluded as a noticeably different PDD, because a delay of this magnitude would be much more likely to be perceived by users as a call attempt that has gone high-and-dry, affecting the perception of quality with respect to call completions rather than PDD.

Now, clearly, rules of thumb like these cannot be hard and fixed, so the analyst must exercise some judgment as to the time differences and percentages to be used in the definitions. For example, if the smallest difference from the last cluster of values included in the normally experienced PDD set were 4.5 s, but the delays 4.5 s or greater represented only 3% of the sample, we might reasonably loosen the criterion for being noticeably different to include 4.5 s. Similarly, the 90% that was exactly achieved by including the PDDs clustered around 4 s in Figure 7.1 (because the example was contrived to be that way) might be less exact in other circumstances, representing something that might be as little as 85%, and still be reasonable as a discriminator of what PDDs are normally experienced.

In other words, the rules stated above for determining what is likely to be perceived as normally experienced and noticeably different PDDs are *heuristic* rather than deterministic, designed to suggest a process by which analysts can begin to characterize predictability of PDDs and assess likely user satisfaction with the expected experience. In fact, in the very likely event that the objective of the analysis of PDDs is to determine whether one circuit-switched service or another will be perceived to be of clearly superior quality with respect to user assessment of their experience with PDDs, the exact numbers used for imputing what will be normally experienced and what will be noticeably different will seldom change the comparison of QoS achieved by comparing the competing services on the basis of four ranked characteristics:

1. Expected proportion of calls for which the PDD will be noticeably longer than normally experienced values;
2. Average amount by which the noticeably longer values exceed the top of the range of normally experienced values;
3. Expected proportion of calls for which the PDD will be noticeably shorter than normally experienced values; and
4. Average amount by which the noticeably shorter values are less than the bottom of the range of normally experienced values.

Given these values, the evaluation of the PDD for the two services then proceeds as a sequential test:

- Is there a significant and substantial difference with respect to variable i, $i =$ 1–4, above?
- If there is, then users are likely to be more satisfied with the service exhibiting the better values.
- If not, then repeat the comparison with respect to variable $i + 1$ above.
- If the list of possible comparisons has been exhausted without surfacing a significant difference, then conclude that there will be no difference in perceived QoS with respect to routing speeds of the two services.

Observe that this process can result in some surprising, but intuitively reasonable results. For example, a service for which the PDDs are clustered around 6 s will be predicted on the whole to be perceived as having better quality than one for which the average PDD is 3 s, but the average is from a distribution with 80% of the PDDs being clustered around 2 s and 20% being clustered around 7 s. The plausible explanation for this implication is that the contrast of 5 s in one call attempt in five is more disconcerting than having to wait an average of 3 s longer on every call attempt.

7.2.4.2 Intrinsic QoS

Not withstanding the inherent subjectivity and imprecision in the users' perception of QoS with respect to routing speeds in circuit-switched services, operations and maintenance personnel need measurements of QoS that will indicate whether the service is performing as well as it should. The objective of evaluation of measures of PDD in this case is determination of how well the PDDs in the currently realized service compare with the best performance that can be expected, given the constraints of the system design, technology, and configuration.

The basic discriminator for such applications is analogous to the normally experienced PDDs defined for evaluation of perceived QoS. However, in this case PDDs distinguished are the *primary route PDDs (PRPDDs)*, represented by the cluster of values about the smallest significant mode in the frequency distribution. In Figure 7.1, for example, the primary route PDDs are those values clustered about 2 s. They have a mean of 2 s and a standard deviation of 0.3, comprise 65% of the sample, and represent the shortest PDDs that can reasonably be expected over the particular origin/destination pair for which they were sampled.

Given the characteristics of the primary route PDDs, the intrinsic quality of the service with respect to routing speed is then characterized by the answer to two questions:

- (Q1) How much does the average PRPDD differ from the best PDDs that can be expected, given the specific configuration of the origin/destination routing system being sampled?

- (Q2) What proportion of the sample lies outside of the portion of the frequency distribution representing the PRPDDs?

Memorandum

To: All skeptics who doubt that users of circuit-switched services accommodate PDDs

From: A bare-foot empiricist

Subject: Results of the proposed experiment

To evaluate the experiment, compare your experience with that described below, and assess the results according to the instructions.

Step (1)

Your guess was accurate to less than 2 s – you are so conscious of PDDs, you must be in the telephone business or a high use activity, such as telemarketing; go to step (2).

Your guess was accurate to 2–3 s – you either have a very good sense of time, or you are indifferent to short variations PDDs; go to step (2).

Your guess was much different from what you observed – how can you doubt that others accommodate PDDs when you are not conscious of them yourself? Your personal experience confirms the hypothesis.

Step (2)

I confidently predict that at least three of the persons asked will initially respond with: "I don't know" "I haven't got the slightest idea" "I've never thought about it" or some equivalent expression of complete indifference. With any luck, one or more will directly confirm the hypothesis by saying something to the effect: "I've never noticed. I'm busy thinking about what I'm going to say".

Your sample includes three or more persons who deny their ability to estimate – the results suggest a strange lack of consciousness of PDDs on the part of persons who have a firm expectation don't they? Go to step (3).

Your sample includes fewer than three persons willing to estimate PDDs – go to step (3).

Step (3)

I predict at this step there was at least one person in your sample who refused to produce a guess even when pressed, perhaps one or two who guessed values that were not too far off from what you observed in step

---

(1), and the rest responded with answers that were gloss underestimates ("immediately", "there is no delay", "fractions of a second"), or over-estimates relative to what you recorded.
Your experiment's results conform to my predictions – now you under-stand, at least, why I claim that absolute PDD is something that users accommodate rather that gauge in evaluating QoS, absent the formal psychological study needed to defend this hypothesis in an academically correct way. Were the delay really important to them, your experiment should have evinced a much higher incidence of *consciousness* of PDD. Your experiment's results do not conform to my predictions – I told you not to try this with telecommunications professionals.
[*Editor's note: We cannot determine whether the author is being serious here or writing tongue-in-cheek, and he refuses to comment. We do know from background research, however, that "academic correctness" has never been his strong suit.*]

The magic number for discriminating PRPDDs is 0.3, which is an estimate of the expected standard deviation of a cluster of values about one of the modes in the frequency distribution of PDDs derived from analysis of literally hundreds of different samples of PDDs. This estimate supports the following algorithm for processing a large sample of PDDs to find a set of values representing the PRPDDs:

1. Sort the PDD values into a list of values in ascending order.
2. Beginning with the smallest sample value, calculate the consecutive differ-ences between samples until a value of 0.3 or greater is detected.
3. Calculate the proportion of values in the list that are less than or equal to the smaller value in the difference located in step (2). If the calculated propor-tion is 10% or greater, then accept the set of values as the PRPDD set.
4. If the proportion is less than 10%, drop the values less than or equal to the smaller value and return to step (2).

Like the process for finding normally experienced PDDs, the process here is heuristic, and the result of its application must be assayed for reasonableness in light of the overall frequency distribution of PDDs. Given a result that appears to be reasonable, however, this process will credibly characterize PRPDDs by both eliminating small parts of the sample that are suspect because the measured PDDs were too short, and identifying a portion of the overall frequency distribution that typifies the fastest PDDs that can be attained over the service that was tested. The proportion of the reduced list of sampled PDD values that are not in the PRPDD set identified in this way thus provides the answer to the second question (Q2) to be answered in evaluating intrinsic QoS with respect to routing speeds.

The remaining question (Q1) is, then, whether the observed average of the PRPDDs is consonant with that expected for calls handled with the fastest routing speeds expected for the circuit-switching system that set up the origin/ destination connection(s) sampled. There are myriad variations in what might be expected, depending on switching system design, technology, and config-uration. However, the principal possibilities can be traced and estimated by reference to the following classification scheme.

Linking. The first set of possibilities to be sorted out are all of the many different types of physical links that might be used in setting up a node-to-node connection through the circuit-switched service. Figure 7.3, for example, displays only 12 of the most likely combinations of links that might be used

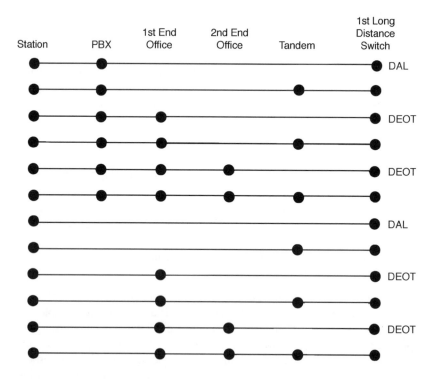

Figure 7.3 Circuit-switched links that might be used for access to a long distance telephone network

to support access to the long distance network from a particular origin station. In addition to the first circuit switch in the long distance network the different circuit-switching facilities displayed there include:

- *PBXs* – private branch exchanges or key sets that manage multiple line/ multiple station terminations;
- *End offices* – the switches used by local service providers to set up local area telephone calls; and
- *Local area tandem switches* – the switches used by local service providers to manage routing of calls into and out of the long distance network(s).

Particular types of links shown there include, for example, lines going directly from the PBX or station to the first long distance switch or from an end office directly to the first long distance switch. In the lexicon of telephony these are referred to as *direct access lines (DALs),* and *direct end office terminations (DEOTs),* respectively.

There are, in addition, an equal number of common linking possibilities for terminations that can be identified by reading Figure 7.3 from right to left, and for any particular connection, the transport may involve from one to as many five or six interconnected long distance switches. This means that there are already as many as 720 different types of end-to-end connections that might be distinguished for purposes of characterizing routing speeds, just in a domestic long distance network, much less for international connections.

Signaling method. For each of the possible combinations of node-to-node links for access and termination suggested in Figure 7.3 the signaling method used to request and set up the interconnect may be either:

- *In-band* – whereby the control information is exchanged over the same link that will be used to carry exchanges of information between the origin and destination; or
- *Out-of-band* – whereby the control information needed to set up an end-to-end connection is carried over a parallel, high-speed data telecommunications system.

Because in-band signaling requires seizure of the interconnecting line and transmission of audible tones that are translated into routing and handling instructions for the switch to which they are transmitted, routing with out-of-band signaling is generally much faster that with in-band signaling.

Registration protocol. The receipt by the distant switch of the routing information for a request for interconnection across the switch is called *registration* of the connection request. The procedure for transmitting that information from the transmitting node to the receiving node is referred to here as the *registration protocol.* While the registration protocols for out-of-band signaling are fixed by the signaling system, and generally involve transmission of completely specified instructions, there are three different registration protocols that are variously used in conjunction with in-band signaling. They are distinguished by whether the transmission of registration data is:

- *Senderized*, so that none of the information elements needed for routing (e.g. digits dialed) is transmitted across the link until all have been collected at the switch sending it and packaged ("senderized") for high-speed transmission;
- *Overlap outpulsed*, so that essential elements of the information needed for routing are transmitted as they become available at the switch sending it; or
- *Cut-through*, so that any registration data is transmitted to the distant switch as soon as it becomes available at the sending switch.

To appreciate the substantial difference in routing speeds among these three protocols, suppose that registration requires in-band transmission of a ten digit North American Numbering Plan telephone number, comprising a three digit area code, a three digit end office exchange number, and a four digit station number. When a user dials out over the direct end office termination using cut-through registration protocol to the long distance switch, the number is registered as soon as the last digit is dialed, so there is no time added to the PDD in setting up the access connection. When the same number is transmitted via a senderized protocol over the same DEOT, the end office does not begin to transmit anything until the last digit is dialed from the user station. When that digit is received, the clock starts on the PDD. The end office initiates transmission of the number to the long distance switch by seizing a line (which takes about 0.5 s) and transmitting the ten digits, which takes about 0.2 s per digit. As a result the PDD is already at about 2.4 s before the ten digits are registered at the long distance switch. In the intermediate case of overlap outpulsing, the end office initiates seizure of a line to the long distance switch and transmission of the digits as soon as the third digit is received, and then transmits three more digits when the sixth digit is received. This way the long distance switch has the six digits needed to determine routing of the call and can execute the onward routing process while the user is still dialing in the last four digits in the number called. Since the clock on the PDD does not start until the last digit is dialed by the user, this process decreases the expected PDD by the time it takes a user to dial four digits, which is usually something of the order of 1.2–2.0 s. Similarly, in cases where the ten digits are mechanically transmitted, overlap outpulsing will reduce the delay in routing a number to about 1.5 s, as compared with the minimum of 2.4 s that would be required were the registration protocol senderized.

Handling at a node. Another factor that will affect the delay in setting up the link between two nodes as an end-to-end connection is whether any additional handling must be executed at a node in order to determine the routing. For circuit-switched services, such handling may include:

- *Translation*, whereby the number dialed by the user is translated into another number that is used for routing throughout the network. For example, in the US, when a free phone number is dialed, there are two translations necessary to route it. First, the local service provider must check the number dialed to determine to which long distance service it must be routed. Then, the long distance service provider must translate the free phone number that was dialed into a regular 10-digit, North American Numbering Plan number for onward routing. Similarly, when seven-digit internal numbers used in virtual private network arrangements are registered with the long distance service provider, the privately defined seven-digit number must be translated to reflect the station address under the public switched network dialing plan for the country in which the destination is located. When translation is necessary, it usually takes the form of a submission of the dialed number or a portion thereof to a computerized translation program that executes a data base look-up to effect the translation and returns the number to be used for onward routing. This process usually adds something of the order of 0.5 s to the routing of a number.

- *Screening*, whereby the number dialed by the user is screened for permission for the user to originate a call of the type requested or to invoke routing restrictions on the connection request. Such screening may take place, for example, in a PBX, to assure that the user's station has permission to originate an international call attempt dialed from the station, or to apply MERS (most economical routing system) criteria to select and specify the long distance carrier to which the call is to be routed. Similarly, screening may be applied to attach a precedence to a connection request, enabling it to pre-empt a line that is already in use for another connection as necessary to assure timely routing, or to verify that the users at the station calling really are willing and able to pay for the charges associated with a caller pays number. The amount of time added to the routing of a particular number by such screening actions is usually small, representing 0.2–0.5 s.

One of the side effects of translation and screening is that whenever either function is based on examination of the full set of digits, and in-band signaling is employed, the registration protocol must be senderized. This means that translation or screening may add much more to the expected PDD than the time it takes to execute the process.

Numbering plan. The final factor that must be considered in estimating the expected PDDs is the *numbering plan* for the service, which determines both the number and variability of digits that must be dialed in order to request a connection. The effects of these attributes on PDD are seen as follows:

- *Number of digits.* When in-band signaling is used, the number of digits that must be transmitted to effect registration at each step in the process determines how long it will take to transmit the routing information from one switch to another. For most public access circuit-switched services, for example, the number of digits required on access is seven for private access arrangements, ten for North American Numbering Plan numbers, and 10–14 for International Numbering Plan numbers. For termination to the destination station, the number of digits required may be four, to address the station from the PBX, or seven, to address the station called from the terminating local service end office, or some number in between. The approximate formula for the time it will take to transmit the required number of digits using in-band signaling is:

$$T \approx 0.5 + 0.1(2n - 1) \tag{23}$$

where T denotes the transmission time in seconds, and n denotes the number of digits to be transmitted.

- *Variability in number of digits.* When the number of digits that must be received in order to route a call is fixed, it is possible to set the digit receiver to scan for the number of digits received and initiate processing as soon as the last digit in a number is received. When the number of digits required to specify the destination station is variable, however, the receiving switch must either execute additional screening to determine how many digits are to be received, or monitor the time between consecutive digits dialed and presume that dialing is complete when an arbitrary threshold for interarrival times of digits has been exceeded. Variability in the number of digits called for in the numbering plan and the means implemented to deal with it can therefore have a substantial impact on the expected PDD. For example, for calling using the North American Numbering Plan in the public switched network, the first digit dialed determines whether the call is a domestic long distance call, a local call, or something else. If the first digit is "1" then the switch sets up to receive ten digits and registration is automatically completed upon receipt of the 10th digit. Similarly, when the first digit is neither "1" nor "0", the system sets up to receive either seven or ten digits, depending on the convention for the local area calling. When the first three digits dialed are "011", indicating that the number following will be for an International Dialing Plan station, the incremental PDD just on the first leg of the access may be increased as much as 3.5–

5.0 s, while the system waits for a period of silence representing the time-out threshold set for presumptive detection of end of dialing. This means that the variability in numbers required can substantially increase the expected PDD in international calls. Whether that increase should be reflected in the estimate of expected PDDs, however, depends on whether it will be realized. For example, the extra delay of 3.5–5.0 s can be circumvented by a caller who knows that "#" appended to the number will be interpreted as an end-of-dial signal. Or, the system may be designed to handle the variability of digits directly by screening the first few digits received to determine the country code or country code/city code pair in the number and use a table look-up to determine how many more digits will constitute a complete number.

Because of such possibilities, it is critical to predicate the estimation of expected PDD on the variability of numbers required to specify a station and at each step in the call set up process, and to determine, or assume, how the number variability will be handled.

7.2.4.3 Expected PDDs

To summarize the preceding descriptions of variations of expected PDDs over segments of the call set up process, Table 7.2 describes some of the possible configurations of links and interconnects, together with their expected contribution to PDD. Table 7.2 can be used to crudely estimate the expected PRPDD for a particular origin/destination connection by determining the configuration of the fastest end-to-end connection supported and adding up the associated delays. The result will be sufficiently discriminating to suggest whether a measured PRPDDs are consonant with the configuration assumed in making the calculations, possibly indicating, for example, that the PRPDDs observed are so much greater than the estimates that there must be some other source of delay not accounted for in the description of the best route.

For example, suppose we have an originating station through a PBX that is connected directly into the long distance network via a direct access line (DAL) calling a distant station that is terminated via a DAL. Then if the registration protocol at the PBX is cut through, Table 7.2 shows that there is no time added, because the last digit dialed is registered directly at the long distance switch. The handling time at the first switch is 0.5 s for translation, because a seven-digit number was dialed. In addition, if the long distance transport utilizes out-of-band signaling there is another 0.5 s for origination of system access. Thereafter, the out-of-band signaling system sets up links at a rate of about 0.2 s per switch. Assuming that the destination switch is the

Table 7.2 Typical step-by-step delays in setting up connections across a circuit-switched telephone network

Link			At-node processing			
Signaling	Registration protocol	Digit Xmit (s)	End-of-digit verification	Add (s)	Other handling	Add (s)
In-band	Cut through	–	Time out delimited	3.5–5[a]	Translation	0.5
	Overlap outpulsed	Seven digits No change			Screening	0.2
		Ten digits 1.7				
	Senderized	Seven digits 1.8	#·Delimited	0.2	Attach ringer[b]	0.5
		Ten digits 2.4				
Out of band		0.2	{Automatic; no end of digit verification required}		Originate out of band signaling[c]	0.5

[a] Depends on time-out criterion set.
[b] Represents only the time to attach the ringer; for PDD a ring signal latency value must be added to estimate the average PDD.
[c] Represents the time required to check into the signaling system, which must be accomplished before downstream routing instructions can be determined and transmitted.

only other long distance switch, the time to set up the call to the distant PBX is then 1.2 s. At the PBX (or distant switch) there is another delay, comprising the 0.5 s worst case for attaching a ringer, and the 1.0 s, or 1.3 s average ring signal latency. Total expected PDD: 2.7–3.0 s. With in-band signaling the use of a senderized protocol on the origination DAL increases the expected PDD by 1.8 s, to 4.5–4.8 s. In-band signaling on the termination DAL then adds another 1.8 s, bringing the expected PDD up to 6.3–6.6 s.

Table 8.2 thus supports estimation of expected PDDs by accumulating delays link-by-link from a description of the structure and characteristics of the end-to-end connection, to develop general ideas of what to expect. It should be noted, however, that the expected delays shown there are nominal values, reflecting the times expected for only a few of myriad possible combinations of equipment, signaling systems, and call handling software that may be in use. They are, therefore, not accurate enough to be used for anything other that their intended purpose here of providing concrete examples of the kinds of variations in PDDs that might be encountered in the global telephone network. For actual evaluation of any particular circuit-switched service, it will be necessary for the analyst to determine from system design or engineering studies exactly what times are associated with the types of links shown for the particular system to be evaluated.

7.2.4.4 Comparisons of Intrinsic QoS

The preceding section describes methods for evaluating intrinsic QoS with respect to routing speeds by determining the best performance and comparing it with expectations. When the objective is to assess *relative* QoS with respect to routing speeds, the measured PDDs can be compared according to the following heuristic criteria, based on samples of PDDs, say A and B, that are large enough to enable:

- Determination of M_A and M_B, the modes in the frequency distribution associated with the PRPDD, and
- Fairly accurate estimation of P_A and P_B, the proportions of PDDs associated with the modes M_A and M_B under the assumption that the standard deviation for PDDs clustered about a mode is 0.3.

As described below these values can be used both for monitoring changes in routing speeds for a particular connection over time, or for comparing two competing services.

Indications of changes over time. In monitoring PDDs for indications of degradation over a specific route, the samples A and B will represent data collected over the same origin/destination connection, using the same service, but at different times. To fix ideas, let A be the earlier time and B the later. Then the measurements will indicate a deterioration in QoS over time when either:

1. P_B exhibits a statistically significant decrease from P_A, or
2. $M_B - M_A \geq 0.6$.

The criteria here assert that either there has been a shift in the proportion of connections being effected over the primary route, or a increase in the mode that would occur as a result of random variations in the measurements less than 10% of the time.

Comparison of two different services. When the objective is to gauge whether there is a difference in QoS with respect to routing speeds between two different services, A and B, then the two samples should comprise measurements of PDDs for either service over the same origin/destination connection(s) taken at the same time, e.g. under a data collection scheme in which the call attempts are interlaced, alternating between the two services. Then results from the two samples can then be compared in accordance with the following heuristic process to determine whether one service is clearly superior to another.

- If $P_A > P_B$, and the difference is statistically significant, and $M_A - M_B < 3$, then service A has superior intrinsic quality. Similarly, if $P_B > P_A$, and the difference is statistically significant, and $M_B - M_A < 3$, then service B has superior intrinsic quality.
- If there is no statistically significant difference between P_A and P_B, but $|M_B - M_A| \geq 3$ then the service with the smaller PRPDD mode has superior intrinsic quality with respect to routing speed.
- If neither condition (1) nor condition (2) demonstrates superiority, then neither service can be ascribed with clearly superior QoS with respect to routing speeds.

7.3 Packet-Switched Services

7.3.1 Concerns

In circuit-switched services the effects of routing speed are visible to the user and manifested only in the establishment of a desired connection. In packet-switched services, each small packet of information exchanged over the connection is individually routed, so that the effects of routing speed are manifested throughout the whole time that the connection is up, in ways that are not necessarily apparent to the user.

The principal concerns with routing speeds in packet switched networks are nonetheless essentially the same as those for circuit-switched networks:

1. How long does it take for a packet to be routed from its origin to its destination?
2. How stable and predictable is the delay time?

7.3.2 Measures

In the language of packet-switching, the measures used to express the answers to these two questions are called, respectively, *latency* and *jitter*, described as follows:

Packet latency. "Latency" is the word used in packet-switching for the time that it takes for a packet to be relayed from an origin to a destination node. The use of the word in this context reflects a perception that unlike information transmitted directly from origin to destination via the connections set up in a circuit-switched service, a packet traversing a packet-switched transport network may be "hiding" somewhere in the network while en route from the origin to destination. As with any store-and-forward relay network, the transmission units comprising packets of information together with their handling overhead are relayed node-to-node across the transport network over links that are dynamically selected step-by-step for each packet. The relay is sometimes accomplished according to routing decisions that "look ahead" and select a number of downstream node-to-node links, but is more generally effected according to route selections that are made after a packet has been received, and is being stored, at nodes traversed during the course of its transmission from origin to destination. This means that the time it takes a packet to traverse an origin-to-destination connection may be substantially longer than the time it would have taken to traverse the same connection in a circuit-switched network at the speed of light. Since the additional time represents the time the packet was being stored and handled at each node, it becomes natural to refer to the origin-to-destination transport time as latency, which may be increased as the packet spends more time hidden at an en route node. Routing speed expressed as packet latency in a packet-switched network therefore has two components:

- *Transmission time*, comprising time required to transmit the packet node-to-node across the network; and
- *Handling time*, comprising the total time that the packet was being held at an intermediate relay node for routing.

Jitter. Because the packets transported across a packet-switched network during the exchanges over a particular origin to destination connection are routed dynamically, the packets do not necessarily traverse the same node-to-node interconnections. This means that any two packets derived from the same block of information to be transmitted may travel over paths that are substantially different with respect to both the number of nodes traversed and the physical distance traveled. Such possibilities for packet-to-packet variations in latency for a set of packets used to exchange information over the same origin/destination connections create what is called jitter in the packet latency. Jitter may be introduced in three ways:

- Variations in transmission times due to differences in the physical distances traveled;
- Variations in handling time due to differences in the number of nodes traversed; and
- Variations in handling time at particular nodes traversed due to differences in the length of queues of packets awaiting routing and transmission at each node.

7.3.3   Quantifiers

Packet latency and jitter are generally transparent to users of packet-switched services. In most cases, their magnitudes are not great enough to be immediately perceptible. Moreover, conditions under which expected latency or jitter might produce perceptible effects are usually mitigated in the design of the packet-switched system, e.g. by use of "jitter buffers", to remove the jitter from streaming data, and precedence routing, to reduce the latency of packets on connections for which the expected inter-packet arrival times might otherwise be great enough to result in lack of activity time-outs on data connections.

As a consequence, packet latency and jitter are measures of intrinsic QoS with respect to routing speeds whose analysis will be analogous to that described earlier for variations in PDDs in circuit-switched services. For this reason, the recommended quantifier for routing speeds in packet-switched networks is *the frequency distribution of values of packet latency sampled for individual packets*. Given such a quantifier we then leave open the possibilities of evaluating intrinsic QoS with respect to routing speeds on the basis of any number of different criteria, such as the:

- Proportion of packet delays that exceed some threshold;
- Probability that the interarrival time between two packets will exceed some threshold;

- Comparison of packet delays for the transmit and receive sides of an origin/ destination connection; and
- "Predictability" of packet delays as evinced by the modes in the frequency distribution, evaluated in the same way as PDDs.

7.4   A Note on Data Acquisition

All of the preceding discussions have been predicated on the assumption of the existence of a data acquisition device capable of accurately recording the time of the last digit dialed (or launch of a packet) and the time of *occurrence* of detection of a particular audible signal (or receipt of a packet). In order to satisfy this condition, any device used for timing events must utilize *zero latency* timing, in which the time of occurrence of any event represents the time it was first detected, rather than the time the event was identified. The best characterization of zero latency timing that I know was made by a long-time friend and colleague of mine, Dr. Peter L. Willson, erstwhile president and CEO of Sotas, Inc., who said:

> In regular timing, we wait to decide what we have heard, and look at our watch. With zero latency timing, we hear something, look at our watch, put our finger on the time, and then try to determine what it is that we are hearing.

This may seem like a trivial distinction, but it is crucial for ensuring the accuracy of what we are attempting to measure when dealing with routing speeds. The most dramatic illustration of the importance of verifying the basis for timing in a data acquisition device occurred when one test unit was replaced with another in one of the large data collection networks with which I have dealt, and I was immediately confronted with the problem of trying to explain to high-level management what miracle of engineering had produced a substantial reduction in PDD across all the origin/destination routes we had been testing. If you want to avoid the extreme discomfort of trying to explain that the data acquisition system that you had been using for years was introducing systemic inaccuracies of which you were unaware, you will be well advised to verify exactly how any data acquisition device you want to use actually records times of events and calculates elapsed time.

8

Connection Reliability

8.1 Evaluative Concepts

The next item on our list of concerns of users of telecommunications services is called "connection reliability", because it is too unwieldy to talk about "connection establishment reliability" or some other adjectives for "reliability" that would resolve the ambiguity as to whether "reliability" in this context refers to the reliability of the connection once it has been established, or the reliability with which connections are established. Since I am, after all, the one paying these words extra by dressing them up in *italics* the first time I use them, I opted to save a little money and ink by letting it be understood that "connection reliability" shall refer to the reliability of the process by which desired connections are established, while the former interpretation, to be covered later, will be referred to as "connection continuity".

User sensitivity to connection reliability in the sense of the term used here, then, is shaped by experience with ordinary telephone service. Nearly everyone has at one time or another experienced conditions under which a series of calls has exhibited a high incidence of failure because of some condition in the public switched telephone network, such as serious equipment failure, malfunction of the routing or switching systems, or inordinately high traffic loads that temporarily exceed the network capacity. Attempts to complete calls under such conditions can be frustrating to the point of exasperation, especially when there is some urgency to the call, such as trying to find out whether someone close was affected by a natural disaster, or trying to contact one's mother before Mother's Day is over.

Because of the sensitivity to the possibility of experiencing a disconcerting lack of connection reliability, users will synthesize their day-to-day use of the telephone, watching for a noticeable increase in incidence of connection failures, and becoming chary any time that a noticeable increase persists, even though problems with connection reliability actually experienced are not anywhere near as vexing as the remembered experience. In this process, users rarely pay attention to, much less fault the service for, sporadically experienced single failures of connection attempts, except when those failures begin to occur with a perceptibly greater frequency. When the occasional connection failure occurs, but the next attempt succeeds, users will almost always attribute the failure to a dialing error on their part, rather than the service. When a user experiences two or more consecutive connection failures, however, the system rather than their dialing becomes suspect, especially when the user has checked the number or consciously dialed more carefully on the second attempt to ensure that the presumed problem on the first attempt was not repeated.

8.2   Concern

All that being said to make it clear what we are talking about, the basic concern of a user of a telecommunications service reflected in connection reliability is expressed in most general terms by the question:

> What is the probability that a correctly executed request for a connection through a network will be extended all the way to the desired destination?

or, more colloquially, by the puzzlement:

> If I do everything right, will my calls go through the first time I dial?

8.3   Measure

By the phrasing of the first question above expressing the user concern, it might seem that the appropriate basis for gauging the QoS with respect to connection reliability of an intermittently-used service is simply P_s, the probability of success of a properly executed request for a connection, as indicated by: (1) set up of the origin/destination connection requested, or (2) determination that the connection cannot be made because the station is already in use. However, because of the relative insensitivity to sporadic single connection failures, the analysis of connection reliability may also require an estimate of P_{f2}, the probability that two or more consecutive connection attempts will result in a failure. Since many of the problems causing failures in connection attempts may substantially increase the probability of a second failure given a first failure, thereby increasing P_{f2} from what would be expected from P_s under the assumption that failures are mutually independent, the most useful generic measure of QoS with respect to connection reliability is the *probability distribution for runs of connection failures*. This probability distribution is defined by a set of values, $\{P_R[i]| \ i = 0, \ 1, \ 2,...,n\}$, for which $P_R[i]$ denotes the probability of occurrence of a run of *exactly* i consecutive failures. In this scheme a run of 0 failures is understood to be a success, so that $P_R[0] = P_s$. The set includes only a finite number of values on the presumption that there will be some value of n for which the probability of experiencing runs of $n + 1$ failures or greater will be too small to show up in a sample unless there is an outage condition. The resultant truncation of the values of $P_R[i]$ to eliminate outcomes that may be theoretically possible, but are unlikely enough to be neglected, then imposes the condition:

$$\sum_{1=0}^{n} P_R[i] = 1 \tag{24}$$

The probability distribution for runs as defined here provides a basis for tests for independence of consecutive failures that is absent when only P_s is used to quantify connection reliability. Whenever is it clear, or can be verified by other means, that the connection failures are independently distributed, the values $P_R[i]$ values can still be readily estimated from P_s by setting:

$$P_R[m] = P_s\big(1 - P_s\big)^m \tag{25}$$

for any value of m and truncating the distribution as necessary to satisfy the condition in Eq. (24) without significant error.

The generic measure of connection reliability defined by the set $\{P_R[i]\}$ satisfying the condition in Eq. (24) is universal; validity of the use of Eq. (25) will depend on the nature of the sample from which P_s is derived.

8.4 Quantifiers

8.4.1 Perceived QoS

To quantify measures of perceived QoS with respect to connection reliability, the estimates of $P_R[i]$ must reflect what users can experience. As described below, the appropriate estimates depends both on whether the service being evaluated is *voice* or *data*, and on whether the objective of the measurement is *characterization* of connection reliability to display expectations, or *monitoring*, to detect occurrence of abnormal conditions.

Table 8.1   User perception of network responses

Normal completions:
>   SBY: *slow (station) busy,* 60 ips (impulses per second), indicating that the
>   destination station is already in use
>   RNG: *ring back signal,* indicating that the connection was extended to the
>   destination station
>   ANS: *station answer*
Failure to connect:
>   RDR: *reorder (network busy) signal,* 120 ips
>   SIT: *special information tone,* a three-tone "warble"
>   RVA: *recorded voice announcement*
>   H/D: *high-and-dry,* no network response after what is perceived to be an
>   inordinate delay

8.4.2.1 Voice

A request for a connection via ordinary voice telephone services is initiated as
a *call attempt,* executed by going off hook, waiting for dial tone, dialing the
station number, and listening for one of the network responses displayed in
Table 8.1. The success of such a call attempt, *so far as the user can perceive it,*
is determined by the network response received. When the response is one of
those shown in Table 8.1 to be associated with normal completions, the user
will deem the call attempt to have been a success, even though there is, from
the viewpoint of the user, rarely any way to determine whether a station busy
signal was received from the station called, or another destination, which, if
known, would represent a failure to properly connect the user's call, or to
verify that a ring-back signal indicates seizure of the right line when the
station called does not answer. Similarly, when the network response is one
of those shown in Table 8.1 to be associated with a failure to connect, the call
attempt will be deemed to be a failure, even though some of the network
responses in this category, such as a reorder signal, may be attributable to
user error, rather than failures in the network and others, such as the high-and-
dry condition, may result from misinterpretation of the PDD, by either the
system or the user.

Characterization. When the objective of the analysis of connection reliability is to assess the likely user perception of quality of voice service in the
long run, the basis for evaluation is the *normal completion rate (NCR)* for call
attempts, defined simply as the proportion of all fully executed user call
attempts resulting in slow busy, ring-no-answer, or answer. The sample of

call attempts from which the NCR is calculated must be carefully qualified, to assure that the assumption of independence required for Eq. (25) is not violated. However, since all call attempts originating from a single site or station are randomized by users' calling patterns, a sample of call attempts will automatically satisfy that assumption, except possibly during brief periods of undetected outage of the origin's access to the network. Thus, almost any sample of call attempts that is distributed over a broad time period will suffice as a basis for estimating NCR and setting $P_s = $ NCR, as necessary to apply Eq. (25) to estimate other values in the set $P_R[i]$.

Monitoring. In some applications the objective of analysis of connection reliability for voice services will be to produce repeated estimates of values of $P_R[i]$ to detect the emergence of problems with completing call attempts. In this case, the possibility of sample-to-sample variations in NCR without any change in the performance of the underlying routing system will generally be too great for meaningful inferences from comparison of the raw NCR values. However, deterioration in performance may nonetheless be clearly signaled by a deviation in the sampled values of $P_R[i]$ for $i > 1$ from the values predicted using Eq. (25) with P_s set equal to $P_R[0]$. Monitoring therefore requires the preservation of the raw data needed to directly estimate runs of failures to effect a normal completion.

8.4.2.2 Data

When the service for which connection reliability is to be analyzed supports intermittently used data connections, such as a dial-up access to the Internet, or transmission of fax or data over the voice network, there are two steps involved in extending the connection request to the destination:

- Set up of a physical, node-to-node connection from the origin to the destination device; and
- Exchange of information between the two devices that enable the exchange of data across the physical connection.

The user perception of connection reliability of attempts to set up end-to-end data connections will, therefore, depend on two measures: the probability that the request for the physical connection will be extended to the desired destination; and the probability of successful set up for the exchange of data across the connection, given a successful physical connection.

The quantification of connection reliability for data services thus requires a supplement to the quantifiers used for voice services. The quantifier needed for this purpose is called here the *handshake success rate (HSR)*, defined by the ratio:

$$HSR = ES/CA \qquad (26)$$

where ES is the number of physical connections over which exchange of *injected* data was initiated; and CA is the number of connection attempts *that were answered* by the destination device.

Thus, for example, in attempts to establish a connection to a distant fax machine or data modem over the public switched voice network, an attempt is counted in CA whenever the appropriate answer tone/synch signal is received at the origin. A connection is not counted in ES until there is evidence of start of transmission of the fax imagery or the data file to be transmitted to/received from the destination device.

Once estimated, HSR is used to adjust the quantifiers of connection reliability for voice services as follows:

Adjustment of NCR. When the quantifier of connection reliability for voice services is the normal completion rate, HSR is incorporated by setting $P_s =$ (NCR)(HSR), the product of the normal completion rate and the handshake success rate.

Adjustment of $\{P_R[i]\}$. When the appropriate quantifier is the full distribution of runs of failures, then each of the values in the set $\{P_R[i]: i = 0,\ldots,n\}$ must be individually adjusted to reflect the stronger requirement for deeming a connection attempt to have been a success. The necessary transform of a set $\{P_R[i]| i = 0,\ldots,n\}$, of run probabilities without consideration of HSR to a set $\{P_{RH}[i]| i = 0,\ldots,n + 1\}$, incorporating the requirement of a handshake success, is achieved iteratively, by setting:

$$P_{RH}[0] = (P_R[0])(HSR)$$

$$P_{RH}[i] = (P_R[i])(HSR)(S_{RH}[i - 1])/(S_R[i - 1])$$

for $0 < i \leq n$; and

$$P_{RH}[n + 1] = S_{RH}[n] \qquad (27)$$

where $S_X[k]$ denotes

$$1 - \sum_{j=0}^{k} P_X[j]$$

8.4.3   Intrinsic QoS

8.4.3.1 Call Completion Rate

For quantifiers of intrinsic QoS with respect to connection reliability, it is necessary to take the user perception out of the picture and define quantifiers that accurately describe what is happening in the handling of call attempts.

The most direct quantifier of connection reliability of this is the *call completion rate (CCR)*, defined as the proportion of mechanically-originated call attempts that will result in a verifiable connection to the station called. The CCR for a service is typically estimated by setting up a test network comprising one or more origin stations and cooperating destination stations that are set aside for testing, or are otherwise configured so that they should not return a station busy signal when called from an origin station at time that does not conflict call originations scheduled from other origins in the test network. A large number of mechanically-dialed call attempts is then originated from test network origins and the network response for each attempt is recorded.

While the data collection for quantifying CCRs is straightforward, the actual calculation of the CCR must be predicated on a careful discrimination of the call attempts in the sample that represent valid tests and those that should be excluded. In particular, there are at least three categories of call attempts that should be eliminated from any sample from which CCR is to be calculated. These are:

1. *Station busy signals (SBY)*. When a test call to a station that is supposed to be free to answer incoming calls returns a station busy signal, there is no way to determine whether there is a conflict in the scheduling of test calls, the station is busy because someone dialed that number by mistake, or the call attempt was, in fact, misrouted to a station that was busy. All test calls for which the network response was a station busy signal should, therefore, be eliminated from the sample.

2. *Blocks of ring-no-answer (RNA)*. When calls to a particular station consistently return a ring-back but there is no answer, it is almost certain that the condition is attributable to the status of the far end test device rather than some failure in the routing system. Since sporadic RNA conditions from a distant device that eventually reverts to a constant RNA probably reflect intermittent failures of the device en route to the complete failure, all call attempts resulting in RNA from a station exhibiting blocks of RNA at some time should be eliminated from the sample. Sporadic, infrequent RNA responses to call attempts otherwise probably represent misroutes by the system and may be included in the overall sample as completion failures.

3. *Patterned consecutive failures*. One of the most important precautions to be taken in screening a sample of mechanically-dialed calls to be used to estimate CCR is to look for and eliminate all but the first call attempt in any group of consecutive failures that exhibits a consistent pattern. For example, if there is a high incidence of consecutive call attempts in which the first call attempt results in a particular failure, say a high-and-dry condition, and the second failure is always a reorder, then it is safe to

assume that the pattern is being created by some systemic fault in the testing rather than some malfunction in the routing system. Any reorders immediately following a high-and-dry condition should, therefore, be eliminated from the sample.

When a sample comprises CA call attempts, the number of call attempts remaining after the adjustments just cited is the count of *valid call attempts*, denoted VCA. Since only call attempts resulting in completion failures have been eliminated in the adjustment, the estimate of CCR from a sample of call attempts containing CS connection successes becomes CS/VCA.

8.4.3.2 Grade of Service

One of the classical quantifiers of intrinsic QoS with respect to connection reliability is the blocking probability ascribed to a trunk group, sometimes called the *design grade of service*. This quantifier is a blocking probability, P_x, defined implicitly in the statement:

> A group of circuits has a grade of service P_x when the expected proportion of call attempts routed to that trunk group during the busiest traffic hour that are blocked because all circuits in the group are busy will be no more than x.

Thus, for example, a P.01 grade of service means that even during the busiest hour no more than 1% of the calls routed to a trunk group will arrive at a time when all of the circuits in the trunk group are busy.

In common discourse the technical notion of "blocking" used in this context is sometimes confused with a "call attempt failure", and the x in P_x is misconstrued to represent $1 - $ CCR. However, there are two major distinctions between the two quantifiers:

1. In addition to the blocking due to lack of facilities reflected in P_x, call completion rates reflect a number of other possible causes of failure to complete a connection, including: sporadic errors in the in-band transmission of dialed digits, resulting in misrouting or inability to extend the call to the destination; sporadic failures of out-of-band signaling systems, resulting in misrouting or a high-and-dry condition; and errors in switching and translation tables, which result in misroutes or erroneous messages to the effect that the call cannot be completed as dialed.
2. P_x refers only to the blocking expected during the busy hour, while a CCR of $1 - x$ reflects the blocking experienced throughout the whole day. Thus, even if blocking due to lack of facilities were the only cause of failure reflected in the CCR, the CCR would be substantially greater than x. Since the busy hour traffic load nominally represents about 30% of the total

traffic, the CCR for connections offering an end-to-end P.01 grade of service would in this case be something like 99.7%.

8.4.3.3 Use-to-Potential Ratio (UPR)

There are some quantifiers of connection reliability whose exact numerical value is not of much use in evaluating intrinsic QoS, but are nonetheless useful when the objective is to maintain the extant level of connection reliability. One such measure is the *use-to-potential ratio (UPR)*, defined as:

$$U_t/P_o \qquad (28)$$

where U_t is the total utilization of the trunk group measured over a 24 h time period, and P_o (standing for "potential") denotes the maximum expected utilization occurring over 24 h when the peak number of active circuits over that time period exactly matches the number of circuits in the trunk group.

More specifically, the utilization, U_t is calculated for a particular trunk group by summing all of the durations of all of the calls carried over the trunk group during a 24 h period, while setting the duration of each call to be the time lapsed between: seizure of the circuit over which it was completed or the beginning of the 24 h period, whichever is later; and the time it was dropped, or the 24 h period ended, whichever happened first.

The potential, P_o, for a 24 h time period is calculated directly from data on carried traffic, much like that which would have been made available at the time that the dedicated trunk group was initially provisioned, showing the start and end times of a large number of calls. From such data, we first produce a curve like that shown in Figure 8.1, showing how the numbers of calls to be connected over the service vary with time of day.The relative numbers of calls expected to be active are then normalized by setting the *y*-axis value of the dotted line in Figure 8.1 equal to 1. The area of the shaded region in Figure 8.1 is then calculated and expressed as proportion, *p*, of the area of the rectangle in which it sits. In Figure 8.1, for example, the total rectangle is 24 square units, while the area of the shaded portion is about 7.7 square units, which gives us:

$$p = 7.7/24 = 0.32$$

The significance of this value of *p* is illustrated as follows. Suppose, for example, that the dedicated trunk group comprised a single T1 termination, which can carry up to 12 voice calls at the same time. Then the total 24 h capacity of that T1 termination would be 17 280 (= 12 calls × 24 h × 60) call

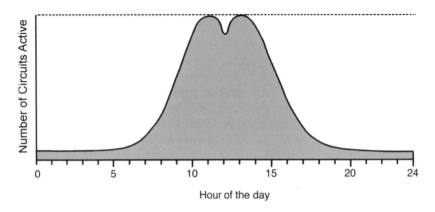

Figure 8.1 Diurnal variations in the number of circuits in use in a trunk group

minutes. However, if the traffic demand varied as shown in Figure 8.1, and the peak value for circuits in use shown to occur at 1100 h and 1300 h in Figure 8.1 were 12, then we would expect to see only 17 280p = 17 280 × 0.32 = about 5530 call minutes carried over that trunk group. And, that 5530 call minutes is, by definition, the potential utilization of that trunk group. It represents the greatest loading in a daily traffic pattern shaped like that shown in Figure 8.1 that can be carried without experiencing blocking because all 12 circuits are busy.

To generalize, then, this means for any trunk group of N circuits, the potential, P_o, in call minutes is given by:

$$P_o = p \times N \times 24 \times 60$$

where p is the proportion calculated in the way just described from data on carried traffic. Thus, a large sample of carried traffic, showing call start and stop times is all that is needed to calculate the necessary potential.

In addition, the data needed to quantify U_t for a particular 24 h period is usually readily available in the form needed for its direct calculation in call detail records, and exactly the value needed is automatically accumulated day-by-day for each trunk group in many commercial switches. This means that once the potential, P_o, for a particular trunk group is calculated from historical data, it is very easy to calculate the UPR on a near daily basis to support monitoring for indications of possible needs to change provisioning of a dedicated trunk group to maintain desired grade of service at the least cost for leased lines.

8.4.3.4 Answer-Seizure Ratio (ASR)

One of the characteristics that makes the UPR particularly useful is that the data required to calculate it is readily available in the call detail records routinely generated for billing purposes. Another quantifier of intrinsic QoS with respect to connection reliability that shares this characteristic is the *answer-seizure ratio (ASR)*, defined generally for given sets of origins (O) and destinations (D) by the ratio:

$$\text{ASR}(O, D) = N_a(O, D)/N_s(O, D) \tag{29}$$

where N_s(O,D) is the number of call attempts to destinations D from origins O, and N_a(O,D) is the number of those call attempts resulting in an answer from a distant station.

This ratio is widely used in the international telephone arena, because the necessary data can be acquired at international gateway switches. For calls routed into a particular country, N_s can be acquired as the number of times lines terminating into that country are seized over some time period, and N_a represents the number of times that answer supervision (a message sent back to indicate that the call attempt was answered and billing should start) was received from a terminating station.

As described in Appendix C, there are a number of pitfalls in trying to calculate ASRs in a way that produces results that can be reliably interpreted for purposes of evaluation of connection reliability. However, as shown in section 8.5.4.2, the ratio can be useful for purposes of monitoring day-to-day traffic for indications of changes in intrinsic connection reliability.

8.3   Evaluation

8.5.1 Assessment of Likely User Perception of Quality

As described at the outset of this section, user perception of QoS with respect to connection reliability will ultimately depend on the perceived risk of encountering disconcerting problems in completing calls to desired destinations, and the perception of that risk depends on deviations from expectations conditioned by whatever is normal for the service with which users are comfortable and familiar. This means that evaluation of perceived QoS with respect to connection reliability must be predicated both on measurements that characterize user expectations and on assessments of what deviations may be noticeable to users.

While the user expectations are best expressed in terms of normal completion rates, which will most closely describe user experience, characterization of user expectations may be based on either the normal completion rate (NCR) or the call completion rate (CCR), as long as it is recognized that there are two major differences between the two quantifiers that must be accounted for in the interpretation of the CCR. The first is that the ratio defining the NCR includes the number of calls resulting in station busy signals in both the numerator and the denominator, whereas the busy signals are eliminated from the sample in calculating CCR. This means that the normal completion rate will always be as large or larger than the call completion rate calculated from the same sample of call attempts. The second is that the normal completion rate experienced by users is usually the completion rate for *manually-dialed* calls, and there is a dramatic difference in completion probabilities between manually- and mechanically-dialed call attempts. In the US domestic network, for example, call completion rates calculated from large samples of automatically-dialed calls typically run somewhere between 99.3 and 99.7%. In large samples of manually-dialed calls over the same routes, the normal completion rate typically runs in the neighborhood of 96.5%! This implies that when we are trying to analyze sensitivity of users accustomed to a service with a 99.5% CCR, we must be sensitive to the fact that an apparently significant deviation from that value may be rendered totally transparent to users by the three calls in a hundred that do not go through because of problems with the accuracy or cadence of their dialing.

Now, when the analysis can be based on comparison of two different services, the evaluation can be predicated on examination of differences to try to determine likely user satisfaction with the service that has not yet been experienced. When such comparisons show, for example, that the service being evaluated has a CCR that is better than or not substantially less than the known service with which it is being compared, we can safely predict that users will be satisfied with the connection reliability of the service being evaluated.

It is, however, an altogether different matter to try to predict at what point users will begin to notice deviations from normal completion rates, much less decide that they constitute too great a risk of disconcerting connection failures. While there are many cases for which deviations from expectations can be credibly judged to be negligible, or clearly deemed "noticeable", the vagaries of human perception, sensitivities and conditioning make it nearly impossible to create a purely analytical means of evaluating user perception of connection reliability, no matter how it is quantified. We can, for example, assess small deviations in CCRs in light of what the normal completion rates tell us about user experience to credibly conclude that that a highly-touted marginal difference of 0.5% in call completion rates will have very little effect on users' perception of connection reliability. Similarly, if we are talking about a domestic service that normally supports an NCR of 96.5%, few would try to gainsay the conclusion that an NCR of 50% "just ain't gonna' hack it", even though that same NCR of 50% might reflect exceptionally good connection reliability in calls into some areas of the world.

It is possible, though, to set some heuristic guidelines for assessing notice-ability of variations in NCRs that begin to clarify the issue. The rule of thumb that I use, without any justification other than it has never failed me, is that:

> Deviations in NCR will be transparent to users as long as the expected incidence of consecutive failures of call attempts remains less than 1–2 per hundred call attempts.

The intuitive justifications for this are that: (1) the users' tendency to discount the first completion failure is both strong and justified by the fact that for modern, domestic public switched networks and isolated connection failure is 5–10 times more likely to be caused by the user rather than the system; and (2) it is difficult to argue that users of a telephone can accurately remember, much less discriminate something that happens on average only once every 50–100 calls.

I am by no means positing this here as something that should be adopted without scrutiny and validation for the particular service being analyzed. However, if you assume that this is a "safe" description of the indifference of users to variations in NCR, and suppose further that consecutive call attempts are mutually independent, then this rule of thumb suggests that normal completion rates as low as about 85% and corresponding CCRs as low as 88–90% may be tolerated by users as satisfactory quality with respect to connection reliability. Thus, even discounting the precise numbers, it can be reasonably surmised that users of *voice services* are much less demanding than the most liberal industry standards for connection reliability.

8.5.2 Assessment of Intrinsic QoS

When the objective of measurement and evaluation of connection reliability is intrinsic, rather than perceived QoS, the evaluation of QoS is a lot more straightforward, but still cannot be accomplished in a vacuum. The basic quantifiers are: the P_x grade-of-service estimates for various linking config-urations, such as access, or end-to-end service; and call completion rates derived from active testing of the service being analyzed. However, none of those values will be meaningful unless there is something against which they are to be compared, such as corresponding values for a competing service, industry standards, or thresholds specified by user representatives trying to acquire a service that they feel comfortable in providing to their user communities. When such bases for comparison are available, the evaluation is accomplished by determining whether there is a statistically significant difference between the measures quantified for the service being analyzed and the values against which they are to be compared. In the absence such bases, there is no way to ascribe any meaning to measurements of intrinsic QoS.

8.5.3 Diagnosis

When the objective of measurement and evaluation of connection reliability is to detect and isolate causes of inferior connection reliability in a network, however, the CCR can become operationally meaningful without an external basis for comparison, as long as internal bases for comparison are built into the sample from which CCR is estimated. In particular, when the test network for actively testing connection reliability in calls placed from a set of origins $\{O_i\}$ to destinations $\{D_j\}$ is configured so that every destination is called by multiple origins and conversely, every origin calls into multiple destinations, then the data from which the overall CCRs are calculated can be arranged into a matrix $\{M_{ij}\}$ like that shown in Table 8.2, in which the values of CS and CA are arranged to show the results for the ith origin calling into the jth destination. The sums across the rows of such a matrix then produce a larger sample for calculating the *origin call completion rate*, reflecting the outcomes of all call attempts originating from the ith origin and the *destination call completion rate*, reflecting the outcomes of all call attempts directed into the jth destination.

As illustrated in Table 8.2, the resultant matrix supports evaluation of the totality of the CCR data to ascribe likely deficiencies in connection reliability by application of the following inferences:

1. If a particular destination call completion rate is low, and there are no significant differences among the origin call completion rates calculated without that destination, then the problem with connection reliability is occurring in the termination of calls to that destination;
2. If a particular origin call completion rate is low, and there are no significant differences among the destination call completion rates calculated without that origin, then the problem with connection reliability is occurring in the access from that origin; and
3. If tests (1) and (2) above fail to identify the likely cause of the problem, then the problem with connection reliability is most likely occurring somewhere in the transport network.

Moreover, in the event that rule (3) applies, it can be further inferred that

Table 8.2 Discrimination of causes of problems in a CCR test matrix

	D_1	D_2	D_3	D_m	$\sum O_i$	$\sum O_{ij}, j \neq 2$
O_1	18/20	17/20	18/20	19/20	72/80 = 0.90	55/60 = 0.917
O_2	16/20	15/20	17/20	18/20	66/80 = 0.825	51/60 = 0.850
O_3	19/20	12/20	13/15	15/20	59/75 = 0.787	47/55 = 0.855
O_n	20/20	14/19	17/20	17/19	68/78 = 0.872	54/59 = 0.915
$\sum D_j$	73/80 = 0.913	58/79 = 0.734	65/75 = 0.867	69/79 = 0.873	265/313 = 0.847	207/234 = 0.885
$\sum D_{ji}, i \neq 3$	57/60 = 0.950	46/59 = 0.780	52/60 = 0.867	54/59 = 0.915	214/238 = 0.899	

where there are more than one origin/destination pair exhibiting transport problems, the first place to look in trying to isolate the problems is on segments of the transport network that are common to the origin/destination pairs for which rule (3) applies. Conversely, when there is but one origin/ destination pair for which rule (3) applies, the first place to look should be any node-to-node interconnections used for that pair that are not, or are infrequently, used in setting up the other origin/destination connections sampled.

For example, in the hypothetical case shown in Table 8.2, we can see that something is awry because the origin call completion rate for O_3 and destination call completion rate for D_2 are both significantly lower than the others, due to a very low call completion rate of 0.60 in cell $i = 3, j = 2$ of the matrix. To discriminate the likely cause, a column is added to the matrix, showing the origin CCRs with D_2 eliminated, and a row is added, showing the destination CCRs with O_3 eliminated. The results clearly point to D_2 as the culprit. This example is contrived (I confess it), but the numbers and results do (I aver) typify real world applications of the discrimination technique.

8.5.4 Monitoring

Where they are needed, evaluations of QoS with respect to connection reliability based on CCR may be the only reasonable basis for comparison of services or rapid diagnosis of problems. However, the requirements for deployment of capabilities for automatically generating call attempts and determining their outcomes severely limits the utility of measurement of CCRs for other purposes requiring connection-by-connection examination of connection reliability. In particular, when the objective of the analysis of a particular service is to monitor connection reliability for indications of deterioration, the breadth of the possibilities makes it practically impossible to deploy and operate enough data acquisition devices to produce anything that even remotely resembles a material contribution to day-to-day maintenance activities.

Effective monitoring of connection reliability therefore requires indicators that can be produced from data that is routinely generated and collected to support network operations, such as the "peg counts" in switches that keep a continual records of utilization of the switch ports, or call detail records (CDRs) for call attempts, which are routinely collected to enable billing for services. As described below, the UPR (use-to-potential ratio) and ASR (answer-seizure ratio) can serve us well in this role.

8.5.4.1 UPR

As described earlier, one of the design standards for the connection reliability of a particular set of trunks or end-to-end connections through a service is the grade-of-service defined by P_x, where x denotes the expected percentage of call attempts during the busiest traffic hour that will be blocked due to lack of facilities. The desired grade of service is achieved in the design of a trunk group by analyzing variations in the loading expected for the set of circuits and ensuring that the number of circuits initially installed is great enough to assure P_x. For example, in setting up dedicated termination into the PBX for a particular customer, the service provider will analyze billing records to determine the peak values on a curve like that shown earlier in Figure 8.1 on the busiest traffic day and throw in a few more circuits as a margin of safety to arrive at the number of circuits between the PBX and service provider switch to carry the expected traffic with a P_x grade of service. The techniques for rigorous estimation of the number of circuits are time-honored and well-proven, so the initial provisioning of the circuits accomplished in this way is usually demonstrated by testing after turn up to achieve the desired grade of service.

There remains, however, the question of how either party will know when the provisioning needs to be changed to maintain the cost-effectiveness achieved in the initial installation. And, the big hitch in this is that the blockage described in P_x applies to *offered traffic*, which includes among other calls directed into the PBX, the outgoing call attempts originated behind the PBX, which will never be seen by the service provider if they are blocked. In practical terms, this means that the service provider has very good data from which to obtain the numerator, C, carried traffic, but has no ready means of knowing the denominator, O, offered traffic, needed to solve the equation:

$$x = 1 - C/O \tag{30}$$

for the value of x that defines the grade of service.

The UPR circumvents this problem by providing an index whose values will signal the possible need for adjustments of the number of circuits. Once the potential P_0 is calculated from the same data as was used in the initial design of the trunk group, the only factor needed to calculate the UPR for any day is the total usage of the trunk group over the 24 h period, which can be readily obtained from peg counts at the provider switch terminating the PBX or from call records.

Given daily values of the UPR the inferences with respect to needs for re-provisioning are straightforward:

1. Values of UPR for the busiest day of the week that are increasing over time and approaching 1.0, indicate that there is an increase in demand that will warrant additional circuits, or the diurnal variations in demand are beginning to deviate from those assumed the last time the trunk group was provisioned. Either way, values of busy day UPRs consistently close to 1.0 signal the need to revisit the design of the trunk group to assure that the desired grade of service is maintained.
2. Values of UPR for busy days that are low or consistently decreasing signal a need to revisit the design of the trunk group to determine whether there are economies to be realized by reducing the number of circuits.

8.5.4.2 ASR

Appendix C describes a number of possible pitfalls in trying to interpret ASR as a quantifier for connection reliability. The positive side of that discussion is a set of criteria that will produce values of ASR that are well enough behaved to serve as fairly reliable indicators of changes in underlying connection reliability, even though the actual numbers are rife with ambiguities. The "trick" is to carefully select the sets of origins $\{O_i\}$ and destinations $\{D_j\}$ for which the ASR is to be sampled and to ensure that each sample is large enough to support reliable interpretation of the results.

The specific requirements are seen as follows. They are predicated on the formulation of the problem of interpreting ASR presented in Appendix C, which posits that the reason ASR might be useful for purposes of evaluation of connection reliability is that ASR may under some conditions represent a stable estimate of P_a, the unconditional probability that a call will be answered, which is mathematically defined by:

$$P_a = (P_c)(P_{a|c}) \tag{31}$$

where P_c denotes the probability that a call attempt will be normally completed; and $P_{a|c}$ denotes the probability that a normally completed call will be answered.

To be used in monitoring for indications of changes in connection reliability, then the estimate of P_a must be derived from ASR values calculated at different times from sample call attempts between origins $\{O\}$ and destinations $\{D\}$ for which:

1. *The sample size is adequate to assure a "tight" estimate of P_a.* Since P_c may be somewhere in the range of 0.90–0.995, the estimate of ASR must be large enough to ascribe statistical significance to differences that are .005 or less. Table 8.2 exemplifies the kind of sample sizes that may be required.

2. *Obvious factors affecting to the values of P_a or P_{alc} have been recognized and controlled in the sample.* Obvious factors affecting P_a include, for example: the mix of business and residential stations reflected in $\{D\}$; time of day; day of the week; and normal business hours in the country called. Obvious factors affecting P_{alc} are numbers of fax and message answering machines. The expected variations due to such factors can be minimized by such devices as:

- Selecting origins so that $\{O\}$ is homogenous with respect to time zone, country, and predominance of types of calls (personal or business), as evidenced by the location of the originating local switch's being in an urban or rural area;
- Similarly selecting destinations so that $\{D\}$ is homogenous with respect to time zone, country and predominance of types of calls; and
- Ensuring that the samples are rigorously screened to assure that samples from times being compared comprise precisely the same mix of hour of day and day of week combinations.

3. *Qualifying any comparisons from one sample to another by assuring that there are no obvious differences in the conditions under which either sample was taken.* This is accomplished, for example, by:

- Assuring that there were no major changes in routing strategy or provisioning of the network(s) carrying the calls from $\{O\}$ to $\{D\}$;
- Verifying that the data from either sample does not appear to reflect some sort of activity, such as testing or number hacking, that creates a large number of seizures that are not intended to result in answer supervision; and
- Verifying that neither sample contains periods, such as national holidays, that would be expected to change the type of calling reflected in the sample.

Table 8.3 Worst-case sample sizes needed for high confidence and accuracy in estimates of ASR

To estimate P_a		An adequate sample size is
Accurate to	With confidence (%) of	
±0.005	95	25000
±0.001	95	43000
±0005	99	620000
±0.001	99	1000000

Given the expected characteristics of the samples from which the values of ASR are to be calculated that will result from attention to these details, the only thing that can invalidate the use of ASR as a means of detecting deterioration or improvement in connection reliability is then a change in P_{alc}, which is inherently a very stable characteristic in sample sizes as large as those contemplated in Table 8.3.

9

Routing Reliability

9.1 Evaluative Concepts

As the perspicacious reader of the previous section will have already noted, the definitions of completions there do not consider the possibility of *misroutes*, which are those instances in which a distant station answers, but the station answering was not the station called. Thus, for example, under the definitions in Table 8.1, any call attempt that is answered is considered to be a normal completion even though the call was routed to the "wrong number". The definitions of quantifiers of connection reliability similarly exclude misroutes from counts of connection failures, except possibly for those test configurations in which the identity of the answering station is testable, e.g. by recognition of a signature response.

There are two reasons for not explicitly recognizing misroutes in analyses of connection reliability. One is practical; the other is theoretical. The practical reason is that there is no easy way to determine whether a connection to the wrong station occurred because of the way that the system routed the call or because there was an error in the dialing. Voice service users can make that determination only when the call is answered, and even then will not know for sure what happened, tending to attribute sporadic misroutes to errors in their dialing rather than in the system's handling of the call attempt. This uncertainty also spills over into data from automatically-dialed call attempts. Without mechanisms that enable the unambiguous identification of the station answering a call, it is nearly impossible to identify misroutes even when the expected station answer is distinctive. For example, even when the expected answer from a station called is a test progress tone, it cannot be readily inferred that a voice answer reflects a misroute when it is impossible to distinguish between a human voice answer from a station, a recorded voice message announcing that the station called is out of service, and a recorded voice announcement to the effect that the call attempt could not be completed as dialed or could not be routed due to congestion in the network.

The practical alternatives, then, are to simply treat misroutes as call attempts that were successfully routed across the network, albeit to the wrong station, or to develop and deploy capabilities for discriminating misroutes from accurately completed calls, which would enable determination of:

1. Whether each number dialed was, in fact, the right number, correctly dialed; and
2. Whether the destination station answering the call was the requested destination, or an acceptable alternative, such as the answer center to which calls for that station are directed when the station does not answer, a different call center programmed to handle overflow from the center called, etc.

Because misroutes are more likely to be perceived by users to be consequences of misdialing and construed to indicate that the system is completing calls, and because misroutes are rarely the cause of completion failures, it makes a lot more sense to count the as perceived connection successes where necessary than to try to distinguish them.

The theoretical reason for treating misroutes as connection successes is equally compelling. When we consider the possible reasons for users' concerns as to how well their telecommunications services handle connection requests, there are two distinctly different perceived risks that may be driving the concern. The first, addressed in the preceding discussion of connection reliability, is the possibility of encountering a situation in which connection failures are frequent enough or persistent enough to become a source of frustration. The associated concern is whether a properly executed connection request will result in an origin/destination connection, or fail en route, which naturally leads to development of quantifiers that accurately reflect ways that the system may fail to set up a connection.

The other perceived risk driving user concerns as to how well the system will handle connection requests is whether information transmitted over a connection will get to the intended recipient. The risk in this case is that misrouting may result in delivery of confidential or sensitive material to the wrong destination. Such consequences are not often contemplated for voice services, because speaker recognition and conversational exchanges usually suffice to assure the caller that the correct station has been reached. However, in any telecommunications service, such as fax or e-mail, in which correct specification of the desired end-to-end connection and the accurate set up of the associated connection are the only guarantors that information transmitted will reach the intended recipient, there is always the nagging question as to whether what was transmitted really got to the person(s) that were supposed to get it, or some unauthorized, perhaps unpleasant or threatening, someone is:

- Having great sport with that painfully mushy Valentine/love letter e-mailed to someone who would appreciate its sentiment;
- Eagerly transcribing account numbers and tracing signatures from that funds transfer request faxed to a stockbroker; or
- Wondering why someone whose name they do not recognize would leave a message on their answering machine to bring $500 cash for bail money, leaving no number to call back.

9.2 Concern

The recognition of the potentially disastrous consequences of misroutes when confirmation of the accuracy of the connection is not possible, then, engenders a general user concern with *routing reliability*, expressed by the question:

> When I request a connection and one is set up, can I be sure that it has been set up to the right destination?

Unlike the concern with connection reliability, which is pretty much limited to intermittently-used voice services, concerns with routing reliability can extend to any telecommunications (or postal) service where there is no means of directly confirming that the destination receiving the information is the one intended.

9.3 Measure

The generic measure of routing reliability is the conditional probability, P_{rlc}, that a connection will be completed to the intended recipient of information or an acceptable alternative, given that it is completed to any recipient.

9.4 Quantifiers

9.4.1 Perceived QoS

From the perspective of the user, there is no possibility of detecting misrouting unless the call attempt is answered. Consequently, the appropriate quantifier for perceived QoS with respect to routing reliability is the direct estimate of $P_{r|c}$, obtained by setting:

$$P_{r|c} = (N_r)/(N_a) \tag{32}$$

where N_a is the number of call attempts answered, and N_r is the number of answered call attempts for which the right station or an acceptable alternative answered.

Since a call must be answered before the information can be passed to any recipient, $P_{r|c}$ can also be estimated indirectly from the answer–seizure ratio (ASR) defined in Eq. (29), calculated from a sample of N_t call attempts and a count, N_m, representing the number of misroutes observed among those N_t call attempts by setting:

$$P_{r|c} = 1 - (N_m)/[(N_t)(ASR)] \tag{33}$$

9.4.2 Intrinsic QoS

While the users' proximate concern with routing reliability is the exposure to having information transmitted to the wrong recipient, which can occur only when the end-to-end connection for transmitting that information is estab-lished, the potential for such an untoward event is created by misdirection that occurs somewhere in the process of setting up a requested end-to-end connection. The quantifier that most appropriately reflects that potential is the unconditional *misroute probability*, P_m, estimated by the ratio CM/CA, where CA denotes the number of call attempts observed in a sample and CM denotes those in which misrouting occurred, regardless of the ultimate disposition of the call attempt.

As described in the discussion of evaluative concepts for routing reliability, there may be no easy way of discriminating misroutes to produce the count CM, even when data acquisition is fully automated. However, wherever an estimate of P_m is required for evaluation, and $P_{r|c}$ is known, P_m can be quanti-fied indirectly by setting:

$$P_m = 1 - P_{r|c} \tag{34}$$

9.5 Evaluation

One of the important reasons for characterizing and analyzing routing relia-bility separately from connection reliability is that users have very little toler-ance for detected misroutes. Users may forgive an occasional brief period of frustration with a temporary inability to complete calls, or tolerate low call completion rates into particular destinations when it is known that there is nothing that the service provider can do about it, for example, because the destination is in a underdeveloped area. However, the first time that a user discovers, even by hearsay, that a correctly addressed fax, e-mail message, or data file was delivered to the wrong recipient due to a fault in the routing system for a service will probably be the last time that service will be described as being "satisfactory", and the advent of pressures on the service provider for assurances that it will *never* happen again!

For this reason, any but negligibly small values of P_m, on the order of 10^{-5} to 10^{-6} or less should be considered to be exhibiting unsatisfactory QoS with respect to routing reliability.

10

Connection Quality – Voice

We have so far measured and evaluated our way *into* a telecommunications service, *through* the time it takes for a connection to be set up, and *onto* an end-to-end connection *with* the right destination. The next natural object of measurement and evaluation is the quality of that connection, as determined by how good voice sounds when it its used for exchanges of conversation, or how good the throughput is when it is used to exchange data. Since these two notions of quality are hardly commensurable (unless we think that all that static stuff we hear when two data modems sync up have something in common with human speech), each is accorded its own separate treatment … and voice quality just happens to be first, because it is the hardest to handle.

10.1 Background

Since the techniques for measurement and evaluation of the quality of voice signals are ones that few people have seen before, while the references to standards like ITU P.800 that might be expected by many readers are nowhere to be found here, "…a decent respect to the opinions of mankind requires that (I) should declare the causes that impel (me) to the separation". Accordingly, I begin here with a brief history of how all this came about.

In the fall of 1980 I was contracted by an embryonic telecommunications company called Satellite Business Systems (SBS) to assist in determining and defining the data collection and analysis capabilities that would be needed to support effective operations and maintenance of their services. At the time, the principal SBS product was envisioned to be private data communications services built around the very high-speed, very low error rate, easily configured links made possible with transport via satellite. Voice services were to be provided, but only as an added inducement to prospective data customers, by offering them a way to use their private data network to avoid expensive long-distance charges for telephone calls.

Because of my background in operations analysis of US Navy telecommunications systems during their transition from terrestrial radio to satellite linked data exchange networks, I was able to describe the necessary system measurement and evaluation schemes in short order, and produce a preliminary design study for an SBS performance analysis system, which shares a lot of words with this book, such as "user concerns", "accessibility", and "quantifiers". My reports were apparently well-received, because SBS recruited me, and in February 1981, I became the nascent Assistant Director for System Measurements, ultimately providing management and technical direction for a small, but highly effective, team of operations analysts. The team was affectionately called the Operations Evaluation Group (OEG), after OEG at the Center for Naval Analyses, where I had cut my teeth on all of this.

There was, however, one small problem. In the preliminary design study I had committed what in retrospect was a tactical blunder, by explicitly declaring for the first time in my career that the performance analysis system would have to include what are described in this book as measures of perceived QoS with respect to various user concerns. The appearance of such a statement in the design study that I authored as a contractor created the false impression that I actually knew how to do this. The presumption was that it would not have been there unless it were, like any other statements from contractors declaring that someone needs something, an advertisement that the something needed is something the contractor knows how to do and is ready to provide on a moment's notice...for the right consideration. Thus, when the shape of the market for SBS began to clarify, revealing that there was not enough demand for data transport to fill the satellite capacity (remember that this was BI, before the Internet with a capital "I"), and commercial voice services appeared to be the only viable filler, I found that I had inadvertently set myself up as the in-house resource, whose consideration was already being paid, for all those someones needing help with analysis of commercial telephone services.

In particular, there came a time when some of those someones expected me to be able to answer a simple question: "If we are transporting voice over digital satellite links that add 0 dB of noise and have 0 dB of signal attenuation, and all of our access and termination interfaces test out good, then why is it that our customers are complaining of poor voice quality?" Well, you might suspect, at the time this happened, I had never even heard of "attenuation" and I wasn't really sure what that dB stuff meant. Consequently, it was hard for me to understand the words, much less what was going on. Moreover, when I looked around for what might be available for evaluation of voice quality, the only ready references were reports subjective user tests (SUTs) that were conducted under a protocol used by AT&T, under which long distance users chosen at random would be asked to rate the quality of calls as being "excellent", "good" "fair" or "poor", answer the question "Did you have any difficulty talking or hearing on this call?", and sometimes provide descriptions of the kinds of impairments that were experienced. I took a quick look at that test protocol, concluding that it was both impossibly difficult to implement for purposes of measurement and evaluation of quality of voice in particular services experiencing difficulties, and deficient in producing results that might be used to establish empirical cause/effect relationships between what a user hears and how that user rates the quality of a call. Thus, I did not even have a very good starting place, much less a vision for how to get there.

What I did have, however, was my experience in the original OEG, working for Navy officers in an environment that did not take kindly to leisurely research, and constantly challenged the analyst to be able to solve problems on the spot. So...what I did this time was what I had been conditioned to do every other time I had been confronted with a problem whose solution required knowledge that I simply did not possess...I made something up. The thing that I made up to circumvent the deficiencies and problems with the SUTs was a voice quality test protocol, whose design was predicated on the following evaluative concepts that guided my formulation of what we had to be able to accomplish if I were going to be able to explain why such an apparently high quality voice service was engendering such distaste.

10.2   Evaluative Concepts

In the context of what is heard over a telephone connection, quality of voice is almost a primitive notion, constituting something that everyone seems to understand, but defying precise definition, because the assessment of quality is inescapably subjective. Expressing the opinion that voice quality is "excellent", for example, is somehow intended to convey in one word an assessment that has been synthesized from the user's experience with the voice service in talking to many different people with different qualities of speech and habits of pronunciation and articulation, under a wide variety of conditions with respect to ambient noise in the vicinity of the hand-set, over connections to many different destinations, completed via many different routes. It is, therefore, almost impossible to ascribe a concrete meaning to such a description. The opinion tells us that the user is probably well satisfied with the telephone service, but it does not convey any information as to *why* the user is satisfied without somehow comparing that voice service to something different enough to change the assessment.

For purposes of such comparisons, there are readily identifiable phenomena that can be heard on a telephone connection whose description is both concrete and unambiguous enough to constitute an objective description of what the user hears. These are *manifestations* of underlying conditions affecting quality of voice that become perceptible through the following process.

A speech signal carried over a telephone connection will comprise two parts:

1. That part which emulates the natural speech of the distant talker with sufficient fidelity to enable the listener to recognize the person talking, understand what is being said, and detect nuances in tone and inflection of speech; and
2. The differences between the emulated speech signal and what is heard.

The differences are created in the processes injecting, encoding, transmitting, decoding, and extracting the talker's natural speech. They are heard as what is "left over" or different once the human listener has extracted from the received signal what is "natural" and expected in the speech of the talker. Since by definition, imperceptible differences of this kind will be assimilated in the extraction of voice signals, what is left over or different is, then, something that is perceptible to the listener. Since that which is perceptible is also distinguishable, recognizable, and describable, this means that there are conditions on a telephone connection whose manifestations can be described for users in such a way as to eliminate nearly all subjectivity in reporting them, even when the users haven't the slightest idea as to what the underlying condition is.

In the jargon of telephony, these manifestations are called *impairments* to the quality of the call. They include, for example:

- *Low (high) volume*, perceived as speech power in the emulated natural speech signal received over a telephone connection that is noticeably less (more) than what would be expected in a face-to-face conversation. (*The distant speaker sounds far away, is hard to hear, sounds too "soft", and is muffled (Too loud!).*)
- *Noise*, heard as audible background signals while the distant taker is silent or left over after the ear has filtered the signal and extracted the emulated natural speech. (*There is static, popping, roar, crackling...on the line.*)
- *Speech distortion*, perceived as qualities or characteristics in the emulated natural speech signal that one would never hear in face-to-face conversations. (*The speech sounds "raspy", "muddy", or "wispy". Sounds like the*

distant speaker is talking underwater or gargling. Sounds strange, weird, warbly.)

- *Echo*, manifested exactly the same way as echo in an empty stadium or a canyon, as a return to the ear of something spoken delayed long enough to be recognized as speech. (*I can hear myself when I talk...talk.*)
- *Cross-talk*, comprising an audible conversation being carried on some other connection. (*I can hear someone else talking on this line.*)
- *Incomplete words*, perceived as gaps in signal power producing missing phonemes or syllables in words recognized in the emulated natural speech signal. (*It sou-s li- some-ing is missing.*)
- *Garbling*, manifested as complete loss of intelligibility in the emulated speech signal, even though the speech signal can still be discriminated by the human listener. (*I cn t ll wu s bng sd.*)

As suggested by these descriptions, sets of impairments can be defined in such a way as to make it very unlikely that a user would be unable to recognize a particular impairment and distinguish it from the others whenever it were present in a telephone call. Such sets of impairments therefore provide a basis for eliciting a description of what a user hears whose only element of subjectivity is the perceived severity or extent of the impairment.

Moreover, as can be readily seen from the examples above, the nature of the different impairments defined can suggest causes that are associated with different aspects of the processing and transmission of voice signals over a telephone connection. For example, reports of cross-talk point to problems of proximity or tuning in analog switching and transmission devices. Similarly, echo is well-known to be created by the "hybrids" that break-out and combine separate transmit and receive signals so that both can be carried on a single line, and is controlled in a network by use of electronic devices such as echo cancelers or suppressors.

What all of this suggests, then, is that any measurements of voice quality of a telephone service must be predicated on subjective assessments elicited in tests in which:

1. Test subjects report not only their subjective opinions of the quality of voice, but also their perceptions of incidence and severity impairments *that are described for them by the person collecting their responses prior to the testing*; and
2. When the service being tested is a new or unfamiliar one, test subjects are exposed to, and asked to report not only on the voice quality of the service being analyzed, but also on the "hidden" samples of calls placed over the service with which they are most familiar.

10.3   Concerns

In the terms just introduced, the basic concerns with quality of voice transmitted over a telecommunications connection expressed by the users' question "How good does it sound?" can be precisely formulated for purposes of evaluation as questions of:

- *Fidelity of the transmitted voice signal.* How naturally and faithfully will the transmitted voice replicate what would be heard in face-to-face conversations?
- *Transmission artifacts.* What are the audible "leftovers" after the best emulated voice signal is discriminated?

The answer to the first question is inescapably subjective, conditioned by each user's auditory filters that enable their cognition of natural speech in sound waves. However, both it and the second question can be answered in quasi-objective terms by describing the expected incidence and severity of impairments and/or their expected effects on the flow of conversation between the stations connected.

10.4   Measures

The industry-accepted generic measure of perceived quality of voice transmitted through a telecommunications service is the subjective description of quality elicited from users of the service, who synthesize their day-to-day experience with its use to assess both how "clean" the connections are and how "clear" the distant speakers sound. To be meaningful, the assessments elicited from users must, I claim, be elicited under conditions specified in section 10.2 in which each user evaluates multiple samples of voice connections. When this is done the associated measures of intrinsic voice quality then become the expected incidence and severity of various possible impairments.

10.5   Service Attribute Tests

The test protocol that I developed to satisfy what I perceived to be the essential elements in a test aimed at meaningful measurement and evaluation of voice quality is the so-called (by me, the caller) *service attribute test (SAT)*. Such a test is conducted by human testers, preferably ones whose only knowledge of telephony is their experience in using it, who place repeated calls to prearranged destination stations answered by persons who similarly don't know much about telephone technology, but are willing to be paid to talk to strangers. The test callers place calls to the cooperating destinations in accordance with schedules that are designed to randomize the sequencing of calls, while ensuring that each test caller makes the same number of calls to each destination over any different services or routes that are included in the test.

The callers are then instructed to hold brief conversations, *without discussing the quality of the call or any apparent impairments*, hang up and report their perceptions and assessments of each call. Data recorded on each call includes:

Perception of impairments. For each call, testers report presence and severity of a set of impairments whose name and nature are defined for them before the test starts. The particular impairments defined for a SAT may vary with the analysts' determination of which are likely to be encountered over the service(s) to be analyzed, and have over the years variously included the ones described earlier, together with refinements, such as identification of different types of noise to distinguish constant noise from impulse noise and noise heard in the background only when the distant person is talking.

The mainstay impairments, included in every SAT since the first one, conducted in the winter of 1981, when I set out for New York City with my pockets packed with money to pay the "call girls", are low volume, noise, speech distortion, and echo. Each impairment in each call is described by the test caller as being:

- *None*, the impairment was not present or was so slight as to have a negligible effect on quality;
- *Much*, the impairment was present, very noticeable, and affected the quality of the call; or
- *Some*, the impairment was noticeable, but sporadic, or otherwise not severe enough to be described as "Much".

In this rating scheme, "none" and "much" conditions are the unambiguous ones, while the provision of the "some" alternative affords the rater the opportunity to describe the presence of an impairment in a call as being neither "none" nor "much".

Assessment of overall effect of impairments noted. Callers are next requested to describe the effect that the impairments reported had on the call as being:

- *None (O)*. There were no impairments noted, or those that were noted were so slight as to have a negligible effect on the quality of the call.
- *Noticeable (N)*. I was conscious of the impairments, but there was otherwise no effect on the quality of the call.
- *Irritating (I)*. I found the impairments to be irritating, but they did not make the call difficult.

- *Difficult (D).* The combined impairments caused me to: raise my voice or ask the called party to speak louder; ask for repetition of, or repeat what was said; change the natural rhythm of my speech; or otherwise actively adapt to the impairments in order to hold a conversation.
- *Unusable (U).* The effect of the impairments was so great that I would have abandoned the connection and re-placed the call, if I were not being paid to make it.

In this reporting scheme, "difficult", "unusable", and "none" are defined by criteria that are straightforward enough to make them quasi-objective, while "noticeable" and "irritating" provide two opportunities for callers to describe situations in which the effect was neither "none" nor so great to satisfy the criteria for "difficult".

This rating scheme was included in the original SAT design with the idea that it would be useful to try to elicit a quasi-objective answer to the question: "How did the impairments reported affect your ability to hear or talk over this connection?" As it turned out, it served to characterize the user perception of "bad" calls in a particularly useful way, by revealing secondary indicators of user satisfaction that are not captured in the subjective assessment of overall voice quality.

Assessment of overall quality of the call. Finally, the callers describe their subjective assessment of the overall quality of the call. These assessments are recorded as numerical opinion scores, on a scale of 0–4, in which 4 denotes "excellent", 3 denotes "good", 2 denotes "fair", 1 denotes "poor", 0 denotes "unacceptable" or "unusable". Half-point scoring is allowed, so that, for example, 3.5 can be used by a caller to grade a call that is better than the calls that have been rated "good", but not quite "excellent", suggesting that the caller was trying to describe the quality as being "very good". Such numbers representing users' subjective assessments of quality are widely known as *opinion scores*. In later versions of opinion scores adopted as an ITU standard, the scale is inflated so that values of 1–5 are used to denote the same responses as the SAT 0–4 values, which we never got around to changing.

In its initial applications in SBS, the SAT proved to be everything we had hoped it would be and more. The convention of using hired callers and cooperating destinations for sampling enabled us to set up a SAT for testing a customer's service quality in a matter of days. Because of the intuitively appealing design of the test, and the fact that we were gauging quality of voice services by the reactions of ordinary users, customers found SAT results to be very credible, and readily acceptable as the gauge of likely user perception voice quality. Part of the reason was that when the SAT results were not good, they invariably confirmed complaints of dissatisfied user communities, and when the results were later turned good, it was usually confirmed that the user complaints had abated.

The real value of the SAT was, however, that its utility as a diagnostic tool was proven time and again, when the distributions of the none/some/much reports for the impairments exhibited patterns that were suggestive of under-lying problems. By reading the patterns of higher than expected incidence of "some" and "much" reports for various combinations of impairments in light of the ways that the impairments manifested might have been produced in the transmission of the voice signals, it became possible to infer the cause of the poor quality with sometimes uncanny accuracy.

10.6   Quantifiers

10.6.1 Perceived Connection Quality

There are two quantifiers of perceived voice quality that can be derived from a sample of calls whose quality is reported by users in SATs, or similarly structured subjective tests of a voice service. The first is the classical *mean opinion score (MOS)*, which is calculated by taking the average of opinion scores. The second is, in SAT terms, $P[\text{UDI}]$, the probability that a call will be rated "unusable", "difficult", or "irritating", as estimated by the proportions of calls in these categories in the samples analyzed.

10.6.2   Intrinsic Connection Quality

Again assuming that the data for measuring voice quality have been derived from an SAT, or similarly structured test, the impairments defined for purposes of the test will provide a basis for characterizing the intrinsic quality of voice connections through the service being analyzed. The quantifier in this case is the *impairments matrix*, showing distribution of reports of the inci-dence and severity of each impairment to produce a profile like that illustrated in Table 10.1.

10.6.3   Evaluation

Data from SATs afford so many opportunities for comparison of the connec-tion quality of two different voice services, the determination of likely causes of user complaints of poor quality, and assessment the likely effects of changes of transmission media or equipment in a voice service that a full discussion of its applications is well beyond the scope of this book. However, there are some

Table 10.1 Typical impairments matrix constructed from the results of a service attribute test

Impairment	None	Some	Much
Low volume	0.901	0.088	0.012
Noise	0.950	0.050	–
Speech distortion	0.983	0.012	0.005
Cross talk	0.997	0.003	–
Echo	0.964	0.028	0.008

caveats and tips that should be kept in mind when trying to interpret SAT-based quantifiers to evaluate voice quality:

(1) *Measurements of perceived QoS with MOS values will be meaningless without a basis for comparison.* There are two reasons for this caveat. First, because an SAT involves repetitious sampling of connection quality and allows for half-point reporting (e.g. 3.5), SAT test callers have a tendency to try to refine their reporting by using a greater proportion of fractional opinion scores than test subjects who are asked to describe at most a few calls with words from a list proffered by the tester. This means that an SAT MOS will tend to be higher than the MOS that would have been obtained from a random sample of callers, and therefore may not be commensurate with the MOS values from other kinds of tests, even when the SAT MOS is increased by 1 to adjust for the difference in scales. Second, as has been cautioned before, but bears repeating – *absolute mean opinion scores are meaningless.* For example, SATs of US domestic long distance services usually produce values on the order of 3.75, corresponding to "very good". If I test a particular service with a group of abnormally critical users whose MOS for US domestic service would be 3.10 without knowledge of that fact, any evaluation of the service tested will be at best worthless and may be totally misleading. This point is emphasized in the dramatization of a real-life experience in the box at the end of this chapter. It is included here with my profoundest apologies to Sir Athur Conan Doyle and all speakers of the Queen's English, whomever and wherever they be.

(2) *MOS and P[UDI] are independent indicators of likely user perception of connection quality.* One of the first things that we discovered when we started working with SATs at SBS was that it was possible for a service to have a high MOS, even relative to its competitors, and still be found to have unsatisfactory voice quality, because of the incidence of "bad" calls. What was happening was that unlike the competing terrestrial services, for which the opinion scores

were usually distributed around a single mode, the quality of voice transmitted via satellite could best be described like that little girl with the curl on her forehead – when it was good, it was very, very good, but when it was bad, it was horrid. Consequently, the MOS for the satellite service represented the average of a large proportion of calls rated as having very high quality and a relatively small proportion of calls exhibiting impairments great enough to have a substantial affect on the use of the connection. The proportion of "bad" calls was, then, not large enough to materially affect the average of the opinion scores reflected in the MOS, but was large enough to create a noticeably higher incidence of calls that users found to be unsatisfactory as reflected in $P[UDI]$.

(3) *Interpretation of impairments matrices is an art, rather than a science.* Although they are displayed as precise numbers in impairments matrices, it must be remembered that the proportions of "none", "some", and "much" like those shown in Table 10.1 are really "fuzzy" values. The imprecision does not, however, impede the evolution of heuristics that can support valid inferences from the totality of results displayed. For example, from the results shown in Table 10.1, I would surmise that there are no problems with the levels in the service tested, because about the right proportion of calls were rated as having "some" low volume. Had the proportion of reports of "none" for low volume been very high, say, 0.99 or more, I would have *suspected* that inbound levels to that listener were significantly higher on average than called for in the loss plan. Had the proportion been substantially lower, say, below 0.80, I would have *suspected* that the inbound levels were low. When I acted on such inferences was I often right? Yes. Are there hard and fast rules that I could have programmed into a computer to make those soft inferences for me? No. How did I arrive at the criteria that prompted me to cite problems? Read the statement in italics above.

(4) *Problems with quality of voice at particular sites can be inferred from MOS values.* Finally, it is worthwhile to point out that when the SATs are conducted from multiple origins to multiple destinations, the likely sources of lower than expected values can be gleaned from analyzing the matrix of origin/destination results in ways similar to those as demonstrated in Table 8.2, showing how to diagnose problems with call completion rates. Specifically, the segregation of results to identify problem routes and attribute likely causes in this case proceeds according to the following algorithm:

1. For each origin in the set of origins $\{O_i: i = 1,...n_1\}$ calculate the average MOS for all calls into a common set of destinations $\{D_j: j = 1,...m_1\}$. Eliminate from the set of origins any for which the MOS is significantly lower than the best MOS values, to produce a reduced set of origins $\{O_k: k = 1,...n_2\}$.

2. For each the destinations $\{D_j\}$ calculate the average MOS in all calls into

the reduced set of origins $\{O_k\}$. Eliminate from the set of destinations, $\{D_j\}$ for any which the MOS is significantly lower than the best MOS values, to produce a reduced set of destinations $\{D_l: l = 1,...m_2\}$.

3. Cycle through steps 1. and 2. using reduced sets until no more origins or destinations get eliminated. When this happens, the set of origins remaining represents a target set that can be used to assess the quality of service into any destination by calculating the average MOS for all calls from that set of origins for each destination. A significantly lower value of the MOS for any destination tested in this way then points to the termination route as the source of quality problems. The set of destinations remaining represents a target set that can be similarly used to test for quality problems in the access side of the connections.

4. Any origin/destination pair that exhibits MOS values significantly lower than the average for all calls between the target origin and target destination sets for which the assessments in 3. did not attribute the problem to access or termination can then be assessed as having quality problems attributable to the transport part of the connection.

One foggy, dreary evening, just as I was about to don my deerstalker and great coat and head for my flat, I was accosted at the doorway by a very nervous engineer, one Mr. E., to whom I had been of some small service in the past. Although I was certain of the reason for his late-hour visit, I bade him to remove his raincoat arid take a seat. The following exchange occurred:

Mr. E: We've got to talk. I...

I: Yes, I know. You desperately need my assistance in interpreting the report we recently sent you. The one detailing the results of subjective testing of the quality of voice over those new echo cancelers. A report, I might add, that gives you exactly the information that you insisted on, despite our advice to the contrary.

Mr. E: Well, yes. That is exactly what I've come about. But how did you know?

I: Elementary. The copy of the report sticking out of your pocket is dog-eared and smudged, exhibiting signs of having been read and re-read in a vain attempt to find a way out of the quandary with which you are now faced. The signs of fatigue about your face and eyes clearly indicate that your searches have not been without some urgency, and the slight tremor in your hands shows that there is some perceived danger should you not be able to obtain directly the answer you now seek.

Mr. E: And if you are such a great detective, suppose, then, that you tell me the rest.

I: Gladly. When you came to us, you asked that we test the new echo cancelers for you, to verify that they do not degrade voice quality. When we tried to suggest what information you would need, you cut us off, and insisted that all you wanted us to do for you was to have test callers try calls completed through the new echo cancelers and produce a mean opinion score (MOS). No other testing was possible with the set-up you had, and you had neither the time nor the money to set up anything better. These facts I am sure you must recall. What I guessed then, and we both know now, however, is that you have exactly the measure you requested, quantified as precisely as you desired, but you cannot decide therefrom whether the new echo cancelers will degrade our users' perception of quality of our voice services.

Mr. E: Yes. And my superiors are pressing me for the decision. I had expected to see something clear-cut. But the MOS value that I have right now is neither fish nor fowl. It is not great enough to clearly indicate that the new echo cancelers improve voice quality, nor is it low enough to allow me to conclude with any confidence that they will harm voice quality.

I: Which, I believe, is exactly the possible untoward outcome that I tried to warn you might eventuate, were we to conduct the testing precisely as you specified. Did I not warn you that your approach was not fully cognizant of the objective?

Mr. E: But the objective was clear-cut. We needed to know what our users' perception of voice quality would be with the new echo cancelers. And that's what I requested.

I: No sir. You just said it yourself. The objective was to decide whether it the new echo cancelers could be deployed without deleterious effect on users' perception of voice quality. And for that you need the additional information that I am now handing you.

Mr. E: What's this?

I: The results from the tests that you didn't ask us for…parallel subjective assessment of voice quality of connections made over circuits with the older echo cancelers by the same test callers. Tests that we added despite your instructions to the contrary.

Mr. E: How on earth did you know I would need them?

I: An elementary deduction, hardly worth comment.

11

Connection Quality - Data

11.1 Evaluative Concepts

For the case of telecommunications connections established for the exchange of data rather than voice, the notions of connection quality are much more precise and objective, depending on predictable effects of signal distortions occurring during the transmission of binary data streams from the origin to the destination. In measuring the QoS over a connection, however, it is necessary to carefully distinguish between:

- *Transmission bit error rates (BERs)*, which comprise differences between the binary data streams transmitted from the origin and what is received at the destination; and
- *Data error rates (DERs)*, which represent the manifestations of BERs as differences between the binary data *injected* at the origin comprising what is to be delivered and the image of that data *extracted* at the destination.

As described in Part I, the process involved here is that the raw injected data, such as a digital data file to be delivered to the destination or the digital scan of an image that is to be transmitted via fax, is transmitted under a *data transmission protocol* that specifies:

1. How the injected data is to be encoded for transmission by addition of other bits to enable detection and correction of errors, control transmission and routing, or describe the source destination and type of data being transmitted;

2. What must be exchanged between the origin and destination in order to set up the end-to-end connection;

3. How receipt, or non-receipt, of the injected data is to be acknowledged by the destination; and

4. If provided, how needs for automatic retransmission of missing or errored data will be recognized and effected by the origin.

The effect of the data transmission protocol is, then, so pervasive that BERs by themselves become totally meaningless as a measure of quality of a data connection, and it becomes impossible even to define quantifiers of quality of a data connection without explicit reference to the protocol used.

Rather, we must take into account that the essential effect of the protocol for a particular transmission medium and its expected BERs can be characterized as achieving a trade-off among three operational characteristics of a data communications service: data error rates, reliability of delivery, and time to complete a transaction. For example, data error rates as a function of bit error rates will be determined by the power and robustness of any error detection and correction coding called for in the protocol, and may be reduced to 0, albeit at the expense of added delay, by use of protocols calling for automatic retransmission of blocks of data received with errors. Similarly, reliability of delivery can be made very close to 1 by use of operational data transmission protocols that require transmission of additional data bits to safeguard against errors in routing and/or additional exchanges supporting verification that the connection has been extended to the proper destination. And, for a fixed bit transmission rate the time it takes to complete a transaction in the face of a particular BER can be minimized by relaxing standards for data error rates or reliability of delivery.

Thus, even though BERs fail as an intrinsic measure of quality of a data connection, there are in its stead three measures of operational performance whose combination can be used to gauge quality with respect to both the BERs that are experienced and the effectiveness of the data transmission protocol used in the face of those BERs.

11.2   Concern

In the final analysis, all users of a data communications service have but one concern that shapes their perception of connection quality. That concern is *transaction time*, expressed in its simplest and most general form by the question:

> Once my data exchange transaction has begun, how long will it take to complete delivery of an acceptably accurate version of the injected data to its destination?

A transaction in this context will be a set of data exchanges that result in delivery of whatever a user views as a logically complete data exchange activity, such as uploading or downloading and complete data file, transmission of all pages in a fax document, or execution of a request and log-on for access to a web page via the Internet.

11.3 Measure

The generic measure of quality of a data connection as perceived by a user coincides with the concern. Since transaction time is the user concern, expected transaction time is the measure of quality.

11.4 Quantifiers

Since the objective of measurement of transaction time is to ascertain how fast various transactions will be completed in under the data transmission protocol employed by the data service being analyzed, there are different types of quantifiers that may be used. As described below, the principal differences that must be taken into account are whether the data service employs:

1. Dedicated or circuit-switched set up, or store-and-forward relay for establishing end-to-end connections; and
2. Fixed- or variable-speed protocols.

11.4.1 Dedicated/Circuit-Switched Set Up

For either dedicated or circuit-switched set up, once an end-to-end connection is established, it is not changed. All the time required to set up the connection is therefore reflected in quantifiers of routing speed, and once the destination device has been verified to be the one called there is no issue of whether part of the data will not be delivered, because it was sent to the wrong destination.

In this context, then, what must be characterized in order to address the users' concern with quality of connections varies with the protocol as follows:

11.4.1.1 Fixed Speed Protocols

Under fixed speed protocols, like a 56 kbps commercial ISDN service or a bulk data transmission network employing 1.5 Mbps node-to-node links, the speed at which data is transmitted is always the same. The rate at which injected data is delivered therefore depends only on the BER and data transmission protocol. For such services, a value $T[b]$, representing the average transaction times for delivery of a block of injected data comprising b binary digits is adequate for characterizing and evaluating the quality of connections. That average may be calculated directly from samples of transmission times for the b injected data bits, or, indirectly, from independent estimates of:

- Throughput efficiency (TE) defined by Eqs. (18) and (19) in section 6.3.3.2;
- Handling overhead (HO) defined by Eq. (20) in section 6.3.3.3; and
- Encoding overhead (EO) defined by Eq. (21) analogous to HO in section 6.3.3.4.

Then if d represents the fixed transmission speed, the average transaction time delivery of b injected data bits can be estimated by setting:

$$T[b] = [b(1 + \text{HO})(1 + \text{EO})]/[(d)(\text{TE})]$$

$$= b/[(d)(\text{TE})]/[(1 + \text{HO})(1 + \text{EO})] \tag{35}$$

Equivalently, the product in the denominator of Eq. (35) can be thought of as the expected *effective data transfer rate (EDR)* achieved under the data transmission protocol over typical connections effected in the service. Since this value can be divided into any number of injected bits, x, to estimate $T[x]$, it is a more versatile quantifier of the actual transaction time. Moreover, since all of the factors in the denominator of Eq. (35) that define the EDR can be estimated as a function of the BER from technical descriptions of the data exchange protocol, the EDR can be used as a quantifier for both intrinsic and perceived connection quality.

11.4.1.2 Variable Speed Protocols

Under variable speed protocols, like those used for transmission of fax and data over ordinary voice telephone links, the data transmission protocols provide for preliminary sampling of the BER on a link and selection of a data transmission speed and format that is expected to maximize the throughput for the BER on the link, thereby effecting a trade-off between expected data error rates and time required to complete a transaction. When such variable speed protocols are employed in a service to be analyzed, there are two equivalent quantifiers for perceived connection quality – the transaction times for a transaction comprising exchange of a fixed number of injected bits, or the speeds at which data was exchanged during the course of completion of a variety of transactions of different sizes, just as suggested above for the case of fixed speed protocols. However, there will in this case be two quantifiers that may be appropriate, depending on whether the data communications transactions are *one-time* or *interactively controlled.*

- *One-time transactions.* For one-time transactions, such as transmission of a document via fax or transmission of a single data file, users will initiate the transaction and then rely on the protocol to handle all further details of the transmission, paying little attention to the precise time involved. Consequently, the average values of transaction times or transmission speeds suffice as a basis for user assessment of connection quality.
- *Interactively controlled transactions.* In contrast, during interactively controlled data transactions, the user executes sequences of possibly related individual transactions, during which the user must remain involved in the activity. This kind of continual involvement, which occurs, for example, when the data communications service is used for interactively controlled transmission of a collection of data files, "browsing" a remote data base by executing sequences of query requests, or two-way exchange of teletype data, creates a consciousness of the possible variations in transaction times that may occur under a variable speed protocol. To satisfactorily characterize transaction times for frequent users of these kinds of transaction, then, it is necessary to use quantifiers that readily communicate the expected variation in transaction times. It is therefore prudent to ensure the greatest generality in the quantifier connection quality by adopting the frequency distribution of EDRs as the standard for any data communications services whose use might be interactively controlled.

11.4.2   Store-and-Forward Relay

As pointed out in previous discussions of other aspects of data communications, when transmission of data is accomplished via store-and-forward relay, the time to complete a delivery of a transmission *unit* from the origin to the destination has two components:

- *Transmission time*, comprising the total time required to transmit the unit over the node-to-node links that effect the end-to-end connection; and
- *Handling time*, comprising the accumulated time that the transmission unit was held at a node in the connection, awaiting routing and onward transmission.

In addition, since the transmitting device has no visibility of the quality of links over which the transmission units are relayed node-to-node, the data transmission protocol must include processes and mechanisms for recognizing when the bit errors experienced during transmission will result in delivery of an unacceptable version of the injected data, and assuring whatever is necessary to ensure delivery of an acceptable version of that data. As with the other types of services, this aspect of management of the quality of the delivered transaction will usually involve provisions for requesting retransmission of any units that are missing or contain unacceptable errors. This means that the trade-offs among data error rates, reliability of delivery, and time to complete a transaction achieved under the data transmission protocol for a store-and-forward relay system will be achieved at the expense of increases in the time to complete a transaction reflected in what can be thought of as *transaction reconstruction time*, comprising the time required to effect all corrective actions required to assure that an acceptable version of the data injected at the origin is extracted at the destination.

As described below, the quantifiers for connection quality in this kind of environment depend principally on whether the transmission unit constitutes an attempted transaction, or only a part thereof.

11.4.2.1 Message Relay

When the data communications service is set up for message relay, the transmission unit is a full message, including the addressing and formatting information needed to control its node-to-node relay from origin to destination. The transaction is delivery of that message to the intended recipient with the injected data, the body of the message, intact enough to accurately communicate to the reader its information content. Since the actual node-to-node transmission time is usually far outweighed by the handling time, the users' first concern with transaction time will focus on expected handling time, as determined by the expected queue delays and precedence handling assigned to the message (if any), rather than the speed of transmission or the size of the message.

In other words, the effects of quality of the connection on transmission are in this case largely transparent to users, and connection quality is manifested in the effects of errors in transmission on the delivery or legibility of the message. In some cases of message relay services, the data exchange protocol (in this case called the message exchange protocol) will provide for message serialization so that recipients can detect when an a entire message is missing, or message tracking, so that an originator will receive an automatic receipt notification when a message was delivered. Beyond such provisions for assurance of delivery, however, it is incumbent on the recipient to review the message text, determine whether something of substance is missing or unreadable, and request that the originator retransmit the message as necessary to complete the transaction.

The appropriate quantifier for connection quality in a message relay service therefore becomes the *retransmission request rate (RRR)*, representing the proportion of messages transmitted via the service for which the intended recipient requests retransmission of all or part of the originally transmitted message because what was received, if anything, did not constitute an acceptably accurate version of the injected data.

11.4.2.2 Packet-Switched Relay

For packet-switched exchanges of data the transmission unit is a packet whose payload represents only a very small portion of the injected data in the transaction. To disguise the fact that "modern" packet-switched networks are but microcosmic versions of message relay networks, the end-to-end delay in transmission of a packet is called the *packet latency*, and the variation in packet latency is referred to as *jitter*. The trade-offs among data error rates, reliability of delivery, and time to complete a transaction are effected by combinations of protocols and hardware. Assurance that all packets are delivered is achieved by dynamic node-to-node routing of each packet. This assures that link failures in the packet-switched network do not prevent establishment of an origin/destination connection, but at the expense of increasing the expected handling time at each node and required size of the buffers at each node to hold packets on queue, awaiting handling. Provisions for increased delivery reliability also increase the transmission time, when the route selected becomes substantially longer than the shortest physical origin/destination connection possible through the network. Data error rates are reduced by increasing the size of the packet to include error detection coding bits and establishing via the protocol what is to be done when a receiving node detects an error in a packet, etc.

Because of all of these various trade-offs, the effects of bit errors incurred in the node-to-node transmission of packets is manifested as an increase in the expected packet latency by the transaction reconstruction time incurred as detected errors are corrected in accordance with the protocol. An adequate quantifier of connection quality is therefore obtained by estimating the ratio:

$$(T_t + T_h)/(T_t + T_h + T_r) \qquad (36)$$

where T_t is the expected transmission time, T_h the expected handling time, and T_r the expected transaction reconstruction time.

Alternatively, if $PL_0[O,D,t]$ denotes the expected packet latency given error free node-to-node links between an origin, O, and a destination, D, under a network traffic load, t, as might be calculated, for example, from network design models, and $PL_A[O,D,t]$ denotes the average packet latency calculated from samples of transmissions of packets from O to D when the network load was t, then the quantifier of quality of connections between O and D through the packet-switched network that will be manifested to users can be estimated by the ratio:

$$PL_0[O, D, t]/PL_A[O, D, t] \tag{37}$$

Another trade-off that may be realized in a packet-switched service is effected by use of "jitter" buffers at the destination node in a connection. The function of these buffers is to hold packets arriving with different latencies in their sequence, so that the transaction bit stream is fed into the receiving device in the proper sequence. When such jitter buffers are used, an out of sequence packet whose delay is different from the least delayed packet by more than the maximum time assumed in the sizing of the jitter buffer is simply dropped because its place in the sequence has already been fed into the receiving device. When jitter buffers are used, then, the increases in the packet latencies due to the transaction reconstruction time, which also increase the jitter, are manifested as increases in the incidence of dropped packets. The *dropped packet rate (DPR)* may then represent an alternative quantifier to those shown in Eqs. (36) and (37).

11.5   Evaluation

It has been posited here that users' proximate concerns with the quality of data connections is always in some form or another focussed on the effects of bit errors on the time it takes to complete a data exchange transaction by transmitting from the origin to the destination an acceptably accurate version of the data injected at the origin. Underlying that concern with the time to complete data exchange transactions is the perception that there is always associated with a given transaction a deadline for its completion. Against such deadlines the expected transaction times will determine how much time a user has in preparing a transaction, and the actual transaction times will determine the probability that the necessary data was delivered to the destination within the budgeted time. This means that:

1. Lower transaction times always enhance the utility of the data service, and
2. Users will always be pressing for lower transaction times for whatever data exchange services they use.

It implies in turn that better values of the quantifiers defined here indicate, without any further interpretation except for verification of statistical significance, a more desirable service.

12

Connection
Continuity

As described in the preceding discussions of connection quality, there may be two distinctly different manifestations of the errors in transmission incurred in a data communications service. The first is a decrease in the speed with which the injected data is transmitted across a connection, as the provisions for automatic error detection and retransmission of transmission units kick in to reduce the number of errors in the data delivered to the destination. The second manifestation of errors, described for store-and-forward relay systems, is an even greater increase in the time to deliver an acceptable version of the injected data, incurred when the provisions in the protocol for reconstruction of parts of a transaction kick in. The difference between these two contributors to transaction times is that the effects on speed of transmission are incurred independently of what is received at the destination, while the additional time required for transaction reconstruction is triggered by information received back from the destination, after the contents of transmission units have been received, extracted, and interpreted.

In the case of message relay, the differences were obvious, because the retransmission requests could only originate after the intended recipient noted from sequence numbers or references that the message was missing, or attempted to read the body of a message received and discovered that parts or attachments were missing, or that it contained indecipherable character strings. In the case of packet switching, the distinction is a little harder to discern. However, were we able to closely examine the transmission of packets, we would see some whose latency was increased because of automatic retransmissions and rechecking in the link-to-link relay of the packet. Then there would be others, whose retransmission was initiated in response to a service message back to the origination, stating in effect that packet so-and-such needs to be retransmitted, because the destination hasn't found it yet, or it was deleted from the system at node x because of detection of uncorrectable errors, because of queue overflow that sent it to bit heaven, or because it was a lowly "discard eligible" packet and the system was just *too* busy to deal with it. The process may produce inordinately long delays in packet delivery, but as long as each destination device can wait long enough, all packets will eventually be delivered in an acceptable form.

In dedicated and circuit-switched data exchange services utilizing the same connection for all parts of the transaction, there is neither an expectation of, nor a provision in the protocol for, the system diligently pounding away until all parts of the transaction are delivered. Instead, a threshold is set for what is reasonable by way of throughput over a particular connection. Then, whenever the connection quality degrades to the point that the throughput drops below a threshold, the protocol gets depressed, gives up, and commits suicide by tearing down the connection.

12.1 Evaluative Concept

The unhappy result of what is sensed by the protocol to be unacceptably poor connection quality is a spontaneous disconnect (*connexum interruptus*) that is manifested to the user in the same way as an outage on one of the links in the connection, by interruption of the transaction in progress. The spontaneous disconnect thus produces a sample of what is described mathematically as an infinite transaction time, which is very hard to average in with their finite cousins.

The impact of such spontaneous disconnects on the user will, moreover, go well beyond any resultant decrease of throughput efficiency reflected in Eq. (19), or any increases in transaction reconstruction time. At the very least, the user will have to repeat the set up of the connection to the destination. Once connected, the user may then have to try to retransmit the whole transaction, rather than just the parts that had not been sent. And, because of the problem experienced with the first transmission attempt, the user may feel the need to "baby-sit" subsequent attempts to complete the transaction, paying a lot more attention to what is going on ("Please, please, don't drop now when there are just another 5000 bytes to go... please?...rats!").

The consequences of spontaneous disconnects experienced by users thus foster a perception of the phenomenon as a different dimension of QoS, even though the proximate causes may be link failures, whose effects are reflected in measures of QoS with respect to accessibility, or poor connection quality, whose effects are reflected in measures of transaction times. This means, for example, that whenever the incidence of disconnects experienced by a user is too great, the service will be deemed to be unsatisfactory, even when the connection quality as measured by transaction times and the accessibility of the service are both satisfactory.

12.2 Concern

The recognition of the possibilities for spontaneous disconnects and experience with their consequences, or envisioned consequences, therefore creates a concern with the *continuity of connections*, expressed by the questions:

Once my transaction has started, will the connection stay up long enough to complete it?, or

Once we start talking, can we keep going until we agree to hang up?

12.3 Measure

The generic measure of connection continuity is simply the probability function, $P[d|x]$, defined as the probability that a transaction, once initiated, will be interrupted by a spontaneous disconnect, given a value of the variable x, which represents some descriptor of the *vulnerability* of a transaction to disconnect, such as the number of bytes to be transmitted or the expected duration of the conversation. The reason for the incorporation of the notion of vulnerability of a transaction in this context is that the longer a connection is up, the greater the possibility of experiencing conditions that result in spontaneous disconnects. We can easily predict, for example, that when there is a systemic problem causing voice connections to disconnect before either party hangs up, the likelihood of experiencing a disconnection will increase with the time the persons tend to talk on the connection. Similarly, all other factors being equal, a 5 page fax transaction is much less likely to be dropped before it is sent than one of 50 pages, and many would agree that the successful transmission of a 500 page fax on the first attempt is right up there with drawing three cards to a straight flush and hitting it.

12.4 Quantifiers

12.4.1 Perceived QoS

An *indicator* of $P[d|x]$ when x is understood to be a whole class of transactions whose vulnerabilities are be averaged out in real world traffic can be derived from analysis of data from very large samples of user reports of disconnects, to produce a *disconnect report rate (DRR)*, defined by the ratio:

$$\text{DRR[T]} = (N_d[T])/(N_b[T]) \tag{38}$$

where T denotes a particular type of transaction (e.g. fax, Internet access, voice conversation), $N_d[T]$ is the number of complaints of disconnects registered for calls of type T, and $N_b[T]$ is the number of billable calls of type T for the same time period and population of users for which $N_d[T]$ was sampled. To be operationally meaningful, estimates of DRR in this way must be derived from:

1. A large data base of customer complaints of disconnects that identifies for each complaint registered: the type of service (e.g. free phone, dial-up public switched, or private virtual network), the type of transaction that was disconnected, and enough other information about the person making the complaint (e.g. number of the station originating the call, city or region from which the call was originated) to define sub-classes of users; and
2. A corresponding data base showing the numbers of connections of different types made over the same time period billed to large homogenous group(s) of customers, such as all residential long-distance service users, or users of a particular virtual private network, whose complaints can be readily distinguished and counted by the type of transaction in 1 above.

12.4.2 Intrinsic QoS

Because of the dependency of the measure on the vulnerability of the transaction, it is very difficult to produce a satisfactory direct quantifier of intrinsic QoS with respect to connection continuity. The best quantifier that can be achieved is through controlled testing to establish a service benchmark. This is accomplished by agreeing on a standard size transaction, such as an *n*-page fax, a *b* byte file transfer, or an *m* minute conversation, conducting tests comprising *t* transaction attempts that were started, and estimating an *abnormal disconnect rate (ADR)*, defined by:

$$\text{ADR} = 1 - (t_c)/(t_s) \tag{39}$$

where t_c is the observed number of the transactions initiated that were completed without interruption and t_s is the total number of transactions sampled.

12.5 Evaluation

When it is possible to conduct benchmark tests of competing services, statistically sound comparisons of ADRs can be used to gauge relative QoS with respect to connection continuity. When customer service activity and billing records generate data bases that satisfy the requirements for meaningful estimation of DRR for different types of transactions, it is possible to compare the values of DRR for different services over time to determine whether connection continuity is improving or degrading.

Beyond this, however, there is not much of an empirical basis for evaluating measures of connection continuity. Absent the opportunity to respond to some sort of outcry from a customer that the overall connection continuity is unsatisfactory, and to document some version of either quantifier for the conditions being experienced, it is difficult to set criteria for deciding when connection continuity is becoming unsatisfactory. I have not seen such an opportunity for more than a decade. This is not to say, however, that connection continuity may not be precisely the problem that must be recognized and tamed the next time that there is a widespread change in signaling or transmission technology.

13

Disconnection Reliability

13.1 Concern

The final item on our list of user concerns that shape the perception of quality of a telecommunications service, is *disconnection reliability*, expressed by the question:

> After I hang up/log off will the connection I was using be taken down in short order?

The basis for this concern is the experience we have all had or heard about in which the disconnect message was not received or received and not properly acted on, leaving the connection up for hours after the user thought it was torn down, resulting in a huge bill for some inordinate number of minutes of service. The effect of such disconnect failures on the users' ability to communicate is so slight that it may seem that this concern is but a formal footnote to what has gone before.

13.2 Evaluation

However, disconnection reliability does serve as very good example of the importance of the "other stuff" discussed in the next chapter, because the users' evaluation of the QoS in this case is not shaped as much by the event itself, as by how the service provider handles the problem for the user when a disconnect failure results in over billing for a call. If the customer service representative readily admits that the time is inordinate, given the user's historical calling pattern, and makes an on the spot adjustment, the user's anxiety with respect to disconnect failures is relieved, and the possibility of others does not represent the originally perceived threat to the user's bank account. If the customer service representative puts up any hint of resistance, or there is any hassle in getting the matter resolved, or the bill is adjusted with a warning that it will be done "… this time, but…", then disconnect reliability can rapidly move to the top of the list of user concerns, completely overshadowing the very high QoS with respect to all of the other factors described here.

In other words, service providers will ignore the possibility of disconnect failures only at their own risk…Not bad for a lowly concern that might just be a "footnote", huh?

14

The Other Stuff

14.1 Evaluative Concepts

In Part I, where the distinctions among intrinsic, perceived, and assessed QoS were drawn, it was posited that where intrinsic quality may make a particular service attractive to a buyer in the first place, intrinsic QoS will be immediately superseded by perceived QoS as the basis for determining whether that buyer will find the service acceptable as it is delivered. Similarly, the perceived QoS that has been the focus of most of this part, though necessary, is not sufficient to assure that users will continue to use it. Rather, that ultimate test of user satisfaction, the loyalty of users, will depend on *assessed* QoS, reflecting users' satisfaction both with the service as it is experienced and with their experience as they are serviced.

The potential effects of user interactions with the service provider as a major determinant of assessed QoS are both concrete and well known. On the negative side, it will be readily granted that the fastest way to assure that a competitor's lines are going to go in where others are now is to have a customer suffer from the negligence, neglect, dishonesty, or rudeness of one of the minions of the providers of those lines. One bad interpersonal interaction is worse than the grievances caused by ten impersonal backhoes...

If there is a caution as to how users of a service with satisfactory perceived QoS should not be dealt with if we are going to keep them as customers, however, there is also an opportunity to deal with them in ways that will make them more tolerant of untoward events that might degrade their perception of QoS. My favorite example of the possibility of such a positive effect comes from the SBS days. Early on in the operation of the SBS system a major customer's multi-node private data network failed and remained completely out of service for more than 56 h. When the service was restored, Operations went beyond the conventional, 'this was an unusual problem, and it should never happen again' report back to the customer, by:

- Presenting a detailed description of all of the emergency maintenance actions that had been taken, and all of the mistakes that had been made;
- Explaining precisely the conditions and deficiencies that had caused the service interruption to last so long; and
- Outlined specific steps that were being taken to assure that the experience would not ever be repeated.

The customer, who could have legitimately been irate and ready to cancel the contract was so impressed by the candor of the presentation and the competence with which the incompetence that had been exhibited was explained that there was hardly a complaint about the outage, much less any move to terminate the service. Such a real world experience demonstrates one of my maxims for fostering satisfactory assessed QoS:

> If there is a problem and you try to minimize it or stonewall it, the customer will never forgive you; if there is a problem and you do everything in your power to make it right as soon as humanly possible, the customer will never forget you.

While maxims like this may be useful for shaping *attitudes* that enhance the likelihood that users' assessment of the QoS will not be degraded by their interactions with the service provider, the more general problem of evaluating the service provider's posture with respect to favorable assessments is more difficult. There are myriad facets of the users' interactions that might shape the users' satisfaction with interactions with the service provider, and the user reactions will be, if anything, even more subjective than the assessment of voice quality, making it nearly impossible to develop measures and quantifiers for all that "other stuff" that shapes assessed QoS.

This is not to say, however, that the evaluation of the effects of user experience in interactions with the service provider cannot be facilitated by the focus on likely user concerns that has guided development of measurement and evaluation schemes for perceived QoS. Rather, what it suggests is that once the concerns are identified, the evaluation of assessed QoS with respect to a particular concern will comprise verification that:

1. The concern is recognized in the day-to-day operations of the service provider;
2. It is the subject of conscientious quality control by the service provider; and
3. Service provider's recognition of, and attention paid, that concern are

clearly communicated to the users in the way that the provider's processes and customer support are characterized for users.

14.2   Typical concerns

The application for this kind of evaluation of likely user assessment of the quality of the service provider is illustrated in Tables 14.1–14.3 showing typical concerns of customers or prospective customers, corresponding possibilities for addressing those concerns in a way that satisfies criterion 3. above, and internal capabilities needed to satisfy criteria 1. and 2. Such profiles provide checklists against which a service provider's posture with respect to assessed QoS can be gauged.

These tables are by no means claimed to be exhaustive, but they do serve to illustrate how recognition of processes and practices that will enhance the likelihood of maintaining customer satisfaction can naturally evolve from articulation of user concerns with the "other stuff".

It should also be noted in this context that the three criteria listed above for evaluating a service provider's posture with respect to concerns other than performance that shape assessed QoS can also be useful in evaluating posture with respect to other user concerns. However, in the case of perceived QoS, it is ill-advised to let the evaluation stop with the examination of posture. The demonstration and communication of the recognition of, and attention to, the users' concerns in terms that users understand them will in this case be a strong selling point, but not nearly as compelling as actually using measures like those described in the previous sections to assess likely user satisfaction and identify steps that might to be taken to improve it.

14.3   Service Level Agreements

The pitfalls in trying rely on what is directly assured users rather than actual measurement and evaluation of perceived QoS is amply illustrated by attempts to substitute actual best efforts with a *service level agreement* (SLA), in which the service provider offers the customer guarantees of service levels that are expressed in terms of measures of intrinsic QoS and provide for monetary penalties when failure to meet those levels is demonstrated. Such an agreement will, for example, be expressed as a contractual commitment whereby the service provider will guarantee that availability of dedicated facilities over a month, measured by some specified standards for identifying outages, measuring their duration, and calculating availability, will exceed a given level. In the event of failure to satisfy the condition

Table 14.1 Provisioning

Customer concerns	What to describe to address customer concerns	To create capabilities for quality control
Installation		
Will the service be installed and brought to full capability	Expected installation process and timing, together with any possible impact on the customer	Develop an operational description of the service installation process and equip sales personnel with capabilities to accurately describe the step-by-step execution and timing of that process
• in a timely manner?		
• with minimum involvement and oversight on my part?	Your track record in meeting delivery schedules	Create and maintain data bases for monitoring step-by-step execution times for handling installations by type of service
Inherent accessibility (dedicated access)		
Will the service as installed meet my perceived needs for reliability and availability?	Specific design service availability, as determined by operational estimates of link availabilities, and any use of redundancy, over-provisioning, or geographic diversity to enhance expected service availability, *rather than* general claims of availability or levels cited in service level agreements	Acquire and maintain information on frequency and duration of outages on links that are used to configure dedicated accesses and terminations
		Develop tools that can be used by sales personnel to demonstrate the effects of configuration on expected availability and price options for achieving specified levels

Table 14.2 Post-installation customer service

Customer concerns	What to describe to address customer concerns	To create capabilities for quality control
Access to service reps (dial-up service)		
• When I have a problem, can I readily get through on a customer service number? • Once connected, how long do I have to wait to talk to someone?	• Expected incidence of busy/fast busy conditions for calls into customer service centers • Expected number of rings until the center will answer • Expected time on hold until a service representative answers	• Establish and monitor a data base displaying numbers of calls into customer service centers by hour of day and day of week • Create and employ models for predicting access delays as a function of provisioning, configuration and call volumes • Create and employ queuing models to predict and control times on hold
Access to service reps (dedicated services)		
• Will I have an assigned account representative/service technician who I can call directly when something goes wrong? • How easy will it be to get to the assigned person or a qualified alternative who knows me when I need to report a problem? • How well will the person assigned know the details of my service and operations?	• Well-defined and clearly specified procedures and standards for support of users of dedicated access services	• Establish a mechanism for reporting back to customers actual performance against the standards for each problem reported

Table 14.2 (*continued*)

Customer concerns	What to describe to address customer concerns	To create capabilities for quality control
Service rep responsiveness		
• Will my reported problems get immediate, undivided attention? • What guarantees will I have that my problems will be worked to the extent necessary to: (a) minimize the deleterious effects on my service; and (b) assure that there is no recurrence • What efforts is my service provider making to: (a) detect possible problems before they degrade my service; and (b) support and facilitate corrective actions?	• Expected time to report a problem and have a trouble ticket opened • Expected time to assignment to a technician, by type of problem and class of service • Expected time to first action by an assigned technician by type of problem and class of service • Standard procedures and practices for problem resolution/ service restoration and follow-up review • Willingness to offer guarantees with penalties for failure to follow these procedures and practices	• Establish reporting criteria for ensuring collection of accurate data showing for each problem times at which: the problem was first detected; the service ticket was opened; a technician was assigned; the problem was diagnosed and isolated; the first technical repair action was taken; the service was restored • Specify and implement a program of performance monitoring in support of proactive maintenance • Establish and maintain formal procedures for assuring that lessons learned from any service action are disseminated throughout the support community

any month, the service provider will rebate some percentage of the charges for that month depending on how much the availability measured is below the guaranteed service level.

When I was first asked to review one these contracts nearly a decade ago, my modestly expressed opinion was that:

Table 14.3   Billing and collections

Customer concerns – are my bills:	How to assert quality control
Comprehensible: In a format that is readily understandable, clearly describing the basis and/or reason for each charge listed	• Establish an ad hoc committee to review the billing formats for ready comprehensibility, usefulness and apparent accuracy Establish a central point of contact for collecting and reviewing questions and confusions with billing formats raised by customers
Useful: Readily analyzable, enabling me to use the data presented to calculate charges under other service a billing arrangements or different conditions, such as a change in expected usage of the service	• Assure that all persons handling billing inquiries are tasked to report any questions of meaning a difficulties in interpreting bills
Accurate: Readily auditable when per use or special condition charges are shown, and free of obvious inaccuracies when I scan the display of charges	• Maintain and monitor a data base of reports of inaccuracies in billing and monitor their incidence for indications of changes following new releases, changes, or corrections to billing algorithms.
Predictable: Arriving at about the same time each month, and reflecting charges over a known, well-defined billing interval	• Set firm mailing dates for bills and use strong incentives for meeting them Spread out mailing dates for different groups of customers to avoid end-of-the-month surges in preparation and mailing of bills
Forgiving: Allowing a reasonable time period for delivery of the bill and receipt of my payment, and demanding penalties for late payment that are neither intimidating nor oppressive	• Review industry practices and ensure that the charges levied by this company, if any, are less than those of the competition Do not turn bill collection activities over to outside companies. Establish an in-house group for bill collections that can be managed to make initial efforts polite and painless

Table 14.3 (*continued*)

Customer concerns – are my bills:	How to assert quality control
Disputable: Supported by readily accessible representatives for handling billing questions and disputes, who: Can be contacted without undue difficulty, such as having to retry may times because the lines are busy, waiting on hold, or having a narrow window during which agents can be contacted Do not constantly re-direct my inquiry to other persons Are knowledgeable enough to competently discuss and understand my problem Capable of rendering a decision and effecting an immediate correction or able to connect me directly with someone with that authority	• Assure accessibility of customer service centers by setting, and monitoring conformance with, standards for number of rings to answer and expected time in queue for call directors • Ensure that persons assigned to handle billing inquiries are thoroughly knowledgeable in formats, billing algorithms, and manifestations of problems, and are trained to recognize legitimate complaints • Empower service centers to adjudicate billing complaints and immediately effect appropriate corrective actions for those complaints that are validated • Upon completion of every call, route the caller to an ARU that will ask whether they ware satisfied with the responses received and automatically route "no" answers to senior personnel to make things right

- The authors did not have the slightest idea of what quality of telecommunications service was all about, because they were expressing service levels in terms of quantifiers of intrinsic, rather than perceived QoS;
- The criteria were expressed in terms of values of quantifiers calculated over a month, when in many cases the monthly estimate of a particular measure could signal a failure for a service whose long-term characteristics are as good or better than those specified by the customer (see, e.g. argument in the box below);
- The criteria were malconceived in the first place, because they had little to do with the way users might be affected by deficiencies in service; and
- All of the criteria for acceptability they defined would be impossibly difficult to routinely monitor to determine when penalties should be invoked.

I therefore concluded that the whole exercise represented the kind of attempt to compensate ignorance with monetary penalties that only a lawyer could logically justify.

But, in those days, I had a tendency to sugar-coat my criticisms.

The triggering condition is that...

...the actual network availability is below the committed network availability...

And the stock reply is...

This is impossible to verify in a timely manner. If you look at the levels of availability the customer is talking about, the MTBFs are 10–20 years. An outage is, therefore, a rare, but possible and conceivable event. much like being dealt a royal flush in a poker game. A single outage at a given site at any time therefore does not necessarily indicate a failure to provide the committed availability even though the availability calculated over any reasonable monitoring time period including that outage would result in a value significantly less than the committed value. In fact, to do so would be the same as unjustifiably concluding that someone were a card cheat because he happened to beat your four nines with a royal flush one night.

Over the intervening years I have frequently been called upon to assist in formulating or evaluating SLAs. While I have occasionally seen some improvements in the formulation of the conditions that will trigger monetary penalties, nothing I have seen has altered my original negative reactions to the concept of an SLA for telecommunications services. That experience includes a series of working sessions carried out as part of a cooperative effort of a customer and a provider to formulate the terms and conditions for an SLA. For this effort I served as the principal, disinterested consultant on QoS, assisting the vendor and customer representatives in formulating unambiguous, easily monitored, and patently reasonable criteria for invoking the penalties under the SLA. If there ever was an opportunity to develop a sensible SLA, this series of joint working sessions was it. Their intent was to identify the concerns of the customer with respect to a wide variety of services and to formulate meaningful definitions of levels of unacceptable quality with respect to each by openly trying to identify points at which a particular problem or shortcoming would become painful to the service users. The working sessions were completely open and non-adversarial, totally guided by enlightened discussions of what would be reasonable for a telecommunications vendor to try to guarantee and for a customer to expect, and devoid of any attempts on the part of either side to cleverly out-maneuver the other.

After four or five all-day working sessions and numerous back-and-forth exchanges of ideas in which the criteria for acceptable quality were clearly and fairly formulated, that particular effort concluded with a mutual agreement between the vendor and customer that the whole idea of the SLA was flawed and should be abandoned in favor of other agreements, without monetary penalties attached, whose enforcement did not require definition in advance of their occurrence of exactly what operating conditions would constitute unacceptably poor quality. Notwithstanding the inherent difficulties in monitoring day-to-day service to detect occurrence of the unacceptable conditions that had been defined, it become apparent to both parties that the concept on which the SLA for the services sought was based was fatally flawed, because *the idea that the inevitable instances of transient poor quality could be reasonably compensated for by what amounted to instantaneous rebates from the vendor was an empty legalistic conception which ignored more reasonable alternatives.*

To understand the sense of this criticism, which led the customer and vendor alike to abandon a good-will effort of formulating a penalty-based SLA, consider for a moment the *de facto* service level agreement that has been evolved for resolving situations when a commercial airline has "overbooked" a particular flight and must leave some of the passengers at the gate behind. The agreement in this case provides for some monetary compensation, usually in the form of a refund of the ticket price and a compensatory travel voucher for any passengers who get "bumped" in violation of the airline contract to honor every reservation. However, to passengers who are anxious to reach their destination on time, a $500 travel voucher for being bumped has very little appeal. At the time they are involuntarily bumped, they are immediately more concerned with what the airline is going to do for them with respect to getting them on another flight (such as booking them first class on a competing airline, if that's the fastest way to get them to their destination). Moreover, after the immediate problem of getting to their destination in a timely fashion is solved, they are likely to remain concerned with the prospect of being bumped again the next time they travel with that airline, and would feel much better eliciting some sort of concession that they would be immune from being bumped from the future flights.

The point, then, is that the travel voucher dictated by the SLA is nice, but other, non-monetary remedies might seem more reasonable and appealing, while the monetary remedy to which the "bumped" customers are entitled begins to some to seem more like protection for the carrier than for the customer. The same applies when an SLA specifies monetary penalties for a period of unacceptable QoS that will inevitably eventuate. In fact, it has been my direct experience in every case of examining SLAs for telecommunications services that the proximate concern of the customer is not "How will I be compensated when something untoward happens?" but

- How can I be sure that you will do everything possible to correct the problem as quickly as possible when unacceptable QoS is experienced?; or
- What kind of leverage will you give me to assure that you are applying your very best efforts to my problems when they occur?; or
- What assurances can you give me that any painful problems we experience will not be likely to recur after they have been corrected?

This experience suggests, then, that like the bumped airline passengers, the proximate concerns of telecommunications service customers are *what the service provider is willing, is able, and will promise to do to make things right when the inevitable unacceptable lapses in service quality do occur.* If this is the case, the idea of establishing monetary penalties for occurrence of problems that have a severe impact on service users becomes one that only a lawyer could love, and customers will be much more interested in negotiating service agreements that specify how the provider shall:

1. React to certain painful conditions, such as a major outage of dedicated terminations, or severe echo being experienced by users of a free phone service; and
2. Assure the customer that any problem encountered has been fully understood and corrected, and that any "lessons learned" will be applied to reduce the probability of recurrence.

Then, if the customer wishes to negotiate further remedies after the monetary penalties of the SLA have been supplanted with promises of such actions, *the monetary penalties, if any, should be severe and imposed on failures of the service provider to act as promised and* not *on the occasional failures of service quality.*

14.4   Quality vs. Economy

Finally, we would be remiss if the consideration of all the "other stuff" here did not to include at least a passing reference to the role played by cost in the users' ultimate determination of the value of a particular service. Although the assessment of cost/effectiveness or cost/utility of telecommunications services is an entirely different matter, well beyond the vision of this book, there are several valuable principles as to how to formulate and analyze the trade-offs between QoS and economy of service (EoS) suggested by the treatment of problems of measurement and evaluation quality herein.

For example, the discussion of assessed QoS in this section clearly shows that service providers should not try to trim costs at the expense of deleterious effects on the factors that determine user assessment of overall QoS. This goes almost without saying, but it does become important enough to warrant the admonition to cost analysts to be sure that any cost/effectiveness trade-off studies explicitly recognize the value to users of the intangible factors besides perceived QoS that determine assessed QoS, stated simply as the principle:

> Attempts to reduce costs at the expense of provisioning, customer service, and billing is penny wisdom and pound foolishness; to ignore what produces assessed QoS is to invite bankruptcy.

Another principle of this kind is:

> The value of an improvement to a telecommunications service must be gauged in terms of perceived, rather than intrinsic, QoS.

In view of the definition of perceived QoS, this may seem like an attempt to belabor the obvious for the unconscious. However, the subtle point here, which has been made several times with respect to relationships between quantifiers for perceived and intrinsic quantifiers of measures of quality, is that there are levels at which further improvements in a measure of intrinsic quality produce little or no improvement in user perception of quality. If these relationships are not recognized in cost/effectiveness analyses, it is entirely possible for a service provider to be encouraged to invest in something, like an increase in call completion rates from 99.0 to 99.5%, that is very attractive from the viewpoint of performance or technology, but has very little impact on user perception of quality.

Finally, for all those involved in analysis of cost and pricing of delivery of service, I offer the ultimate heresy that the never-ending argument between advocates of economy and quality may, in fact, be a totally meaningless one, because:

> It is possible to improve perceived quality with no increase in the cost of delivery of service.

The message here is this. In evaluating alternatives for telecommunications services, it is usually presumed, if not explicitly assumed, that there must be a trade-off between EoS and QoS. Such presumptions lead us to expect, for example, that: investments in quality improvements must be justified by an expectation that better quality will attract and retain more users, or warrant a higher price for the service; a lower QoS will be less attractive to the user community and must therefore be delivered at a lower cost and price if the service is to be competitive; etc.

While such trade-offs between economy and quality of telecommunications services probably do account for much of the telecommunications market mechanisms, it is also true that there are many cases in which QoS and EoS are, in fact, not in conflict, and may actually be complementary, in the sense that there are service delivery options that simultaneously enhance both QoS and EoS. This means that there are opportunities in the telecommunications world to improve quality with no increase in cost, to reduce costs with no loss of quality, or even to improve cost and quality at the same time.

The truth of this observation is best seen from an example that readers can test for themselves. Consider one of the big problems assessing the presumed trade-off between quality that is confronted in deciding how best to provide for those conditions when additional capacity or alternative routes outside of the provider's own network are needed to deliver traffic. For any particular destination there may be many different alternative carriers, each offering a different unit price for use of their networks and facilities. Because the lower price alternatives also tend to offer poorer performance, there is always a concern that the "obvious" solution of using the lowest priced routes will materially degrade user perception of quality, with dire consequences in network management centers and/or the marketplace. As a consequence, there tend to be on-going arguments between service provisioning and finance as to whether cost or quality will be the determining factor in the selection of providers of alternate or overflow capacity. When major customers begin to complain about the QoS after the bargain-of-the-month reseller's trunks are moved to first choice in the overflow routing plan, the operators will readily cite that problem as clear evidence of the fallacy of the lowest cost strategy; when use of the bargain-of-the-month has no apparent impact on user satisfaction, the advocates of the lowest cost strategy are quick to latch onto that experience as evidence that service does not have to be "gold-plated".

To see how the adoption of the appropriate evaluate concepts and measures might allow for an alternative that is satisfactory to both the finance and provisioning personnel who are at loggerheads over the choice of strategy, pretend for a moment that we are they, and suppose that there is some destination, D, for which we expect to be offered substantially more traffic than our network can carry. Suppose, further, that it has been decided that simply blocking the excess is not an option, because our customers reasonably expect a better grade of service to D. In this circumstance it is tempting to conclude that our objective is to locate sources of enough extra capacity to D to ensure an adequate grade of service, and selecting from the alternatives identified the source(s) which will carry the traffic to D that we expect to hand off for the least cost.

In other words, it is tempting to posit that the objective is to procure the minimum amount of extra-network capacity needed to assure adequate QoS with respect to handling the expected traffic to D, and do so at the least cost. However, it is more useful, and probably more accurate, to posit that the objective here is actually to *realize the greatest income from the potential revenue represented by the offered traffic to that destination.*

This objective, then, immediately suggests that the criterion for selecting among alternate sellers of capacity to destination D should be based neither on a measures of QoS nor costs. Rather, the appropriate measure is *expected return ratio (ERR)*, which can be defined generically as the ratio:

(amount of revenue expected from a call overflowed to D)

/(cost of providing for the overflow)

A pretty good quantifier for this measure can be defined as the ratio:

$$[(ASR)(BD)(PC)]/\{(CST)[(1 - ASR)(PDD + UT)$$

$$+(ASR)(PDD + AT + BD)]\}$$

where ASR is the answer seizure ratio; PC is the price per minute of conversation charged to the customer for a completed call to destination D; BD is the average billable duration of an answered call to destination D in minutes; CST is the cost per minute of use of the overflow route; PDD is the average PDD for calls completed via the alternate route; AT is the average time to answer for calls answered at destination D; and UT is the average ring time for calls not answered.

Now, if we look at the factors in this quantifier of the expected return ratio, PC, BD, AT, and UT will be stable and fixed for calls into destination D. The other three will be characteristics that may vary from seller to seller. And, note in particular that CST is the measure advocated by the "least cost" strategists, while PDD and ASR are two of the measures most frequently cited as bases for the criteria advocated by the "best quality" strategists.

I could at this juncture, then, go on to concoct examples of situations where analysis of this ratio for competing sources of overflow capacity would alternately favor the lowest cost source, the source offering the best ASR and PDD, or some compromise in between. However, I think I'll leave that exercise to the curious reader, and be content with the obvious conclusion from the definition of this quantifier of ERR that since CST, PDD, and ASR are independent variables, it is entirely possible that an analysis of alternatives will show that the source that would have been chosen on the basis of least cost may also be the one that would have been chosen on the basis of the best values of PDD and ASR...

Afterword

Since Part I of this book ended with a description of the concerns of persons who use telecommunications services, it is somehow perversely fitting that Part II ends with a similar description and discussion of the concerns of those who have to deal with the service providers in paying for telecommunications services and assuring that the quality of what is paid for is satisfactory. If the intent of these descriptions of such concerns has been realized, readers will have found them to be so intuitively credible as to be self-evident. The user concerns in Part I, for example, should be immediately recognized and appreciated by anyone who has ever used a telephone, absent any knowledge or understanding of telecommunications technology. The "other stuff" described at the end of Part II should be readily recognized and appreciated by nearly anyone who has had experience in dealing with a telephone company, independent of any knowledge or understanding of the sophisticated management theories, organizational concepts, processes, procedures, and policies that determine how the service providers interact with their customers.

Description of those readily apprehensible concerns in terms devoid of technical language may, in fact, have smacked to some as "belaboring the obvious" or "unnecessarily tutorial". Yet, I dare say that few would argue with the premise that these simple, concrete, kindergarten concepts have served us well in the effort to characterize quality of service (QoS) for telecommunications, illuminating and motivating definitions and derivations that might otherwise have been nightmarishly obscure. To the extent that the reader has found this to be true, this book conveys by demonstration the message that the key to credible, cost-effective, scientifically defensible measurement and evaluation of QoS is the preliminary characterization of what is, or may be, important to those who will ultimately determine whether perceived and assessed quality will be acceptable, expressed in terms that are meaningful to those who will be making the judgments.

Such a characterization is, in technical terms borrowed from philosophy, an ontological model of the service being analyzed. I have on occasion tried to stress the importance of such ontological models in analysis of QoS by appeal to the maxim, obviously formulated on the premise that the worse the pun, the greater the likelihood that it will be remembered, that:

If you want meaningful measures of quality of service, ask not what de tech would use; ask rather what the user will detect.

It has been my experience in nearly 35 years of defining measures and quantitative evaluation schemes that analyses predicated on a good ontological model proceed almost unerringly to operationally meaningful measures, easily calculated quantifiers, and evaluation criteria that are credible to users, readily acceptable to decision-makers, and accurately predict the likelihood of satisfaction with perceived quality. The usefulness of such analyses has, moreover, survived the competition (or non-competition, depending on one's viewpoint) from such proffered replacements for the discovery of truth as mathematical programming, logit regression, expert systems, "data mining" and neural networks.

It is such experience, clearly evinced, to some extent at least, in what has been presented here in Part II, that returns us to where we began in Part I, to my ontological model of analysis and the corollary admonition to begin each new QoS analysis effort with something that cannot be implemented on a computer, no matter how fancy and colorful the graphical user interface – a trip into the minds of the persons who will be assessing quality, to determine what they will experience and how those experiences will shape their concerns...

...or so it says here in fine print.

Appendix A

An Example of Formulation of an Analysis

Evaluative Concepts And Measures For On-Call Provisioning

A.1 Introduction

As used in this example, the term "on-call provisioning" refers to any tele-communications service in which the provider establishes temporary, dedicated circuits between specified sites in response to user requests. The objective is to define intuitively credible measures of percieved quality of such servcies. The presentation is divided into two major sections:

- *Evaluative concepts*, which describe user concerns with on-call provisioning, and define generic measures of performance/effectiveness with respect to those concerns; and
- *Quantifiers*, which define quantifiers for the generic measures.

A.2 Evaluative Concepts

The principal attraction of on-call provisioning services is that they afford an alternative to costly overbuilding to achieve resiliency to transient surges in demand or losses of capacity in private networks. As described below, the principal determinants of whether a particular service will be attractive in this application are, in turn:

- *Responsiveness* to expected needs for temporary augmentation of capacity; and
- *Expected cost* of the service relative to alternatives for achieving the required resiliency to transient conditions.

A.2.1 Responsiveness

Since the principal objective in acquiring on-call provisioning services is to circumvent the deleterious effects of transient network problems, one of the major concerns of prospective customers will be whether the service can be relied upon to provide requested capacity soon enough to avoid substantial impact on the activities of users of their networks. The appropriate generic measure of effectiveness with respect to avoiding such substantial impacts is the responsiveness of an on-call provisioning service, expressed as:

PS = the proportion of requests for capacity that will be met in time

to avoid major problems

The value of this measure for any particular service, customer, and class of possible requirements for on-call provisioning will, in general, depend on three factors:

- *Provider response time (PRT)*: the time it takes the service provider to set up on-call circuits in response to requests.
- *Lead time (LT)*: the amount of advance notice of the need that is given the provider, as represented by the difference between the time the request is received and the requested capacity is desired/needed.
- *Shortfall tolerance time (STT)*: The period of inherent resiliency to the shortfall to be corrected, as measured by the time it can persist before the problem is likely to produce noticeable effects on activities of users of the network.

In terms of these times, PS is estimated as the proportion of requests for an on-call provisioning service for which:

PRT < LT + STT

where the term LT is understood to denote lead time if it is positive, and lag time, defined by the latency of the problem creating the need for additional capacity if it is negative.

Factors that affect PRT, LT, and STT, and therefore affect the overall value for PS are:

A.2.1.1 PRT

As described later, the principal determinants of the provider response times are the procedures followed in preparing for, and responding to, user requests for on-call provisioning. The range of procedures that might be employed is, in turn, constrained by: (1) the availability, or lack of, support capabilities; (2) performance of the resources used for circuit turn-up; and (3) expected performance of the provider facilities used to achieve the requested capacity.

A.2.1.2 LT

The amount of advance notice that can be reasonably expected from the customer will vary principally with the nature of the needs for extra capacity. Some needs will result from planned user actions, such as one-time telemarketing efforts, periodic exchanges of very large data bases, or network reconfiguration actions. Provisions for satisfying these needs can be scheduled well in advance of their emergence. Other needs, such as those arising from system failures, may emerge very quickly, and without warning, leaving no time for advance notice. The actual lead time, then, depends both on the potential lead time afforded by the nature of the shortfall the user wants to correct and any delays in notifying the provider when the need is known.

A.2.1.3 STT

The tolerable period of duration of a shortfall varies with the nature of uses of the private network affected by the shortfall. When recognized, some unscheduled needs for on-call circuitry, such as those created by failures of communications links supporting on-line interactions with a remote computer at the height of the business day, may require a very rapid response to avoid substantial impacts. Others, such as similar failures reducing store-and-forward data exchange capacity during a slow traffic period, may pose less stringent requirements for timeliness. The notion of shortfall tolerance time introduced here recognizes the possibility for such variations, and provides a basis for including them in the measurement of effectiveness of on-call provisioning.

A.2.2   Expected Costs

Assuming that the responsiveness of an on-call provisioning service is adequate for the purposes of a user, the critical factors in the decision to buy a particular service will be:

1. The cost of that service relative to the alternative of designing into the private network enough surplus capacity, redundancy, diversity, etc. to achieve the same resiliency; and
2. The expected costs of that service on an event-usage basis.

A.2.2.1 Costs Relative to Alternatives

Because the long-distance transport capacity provided through on-call provisioning services is sold on the basis of use, rather than availability, and the expected use is infrequent, prospective customers expect to have to pay both premiums and premium usage charges for the on-call provisioning service. In addition, there may be some associated one-time or fixed costs to the user, in the form of:

1. Installation and monthly charges for access and egress facilities enabling the user sites to be protected with the provider's network;
2. Features or enhancements of customer premise equipment that may be required by the provider to support provider monitoring of access and egress facilities; or
3. Costs of procuring and operating capabilities to communicate with on-call provisioning centers/reservation programs required by the provider.

Despite such additional costs associated with use of on-call provisioning, where the private network designers take into consideration the possibility of using on-call provisioning at the outset, they will generally find that the architecture that minimizes costs of achieving the desired level of resiliency is a mix of dedicated and on-call facilities.

A.2.2.2 Expected Costs Per Use Event

However, for existing private networks, in which the extant level of resiliency has been set by design, the mix of dedicated and on-call facilities that might be achieved by access to on-call provisioning is likely to be sub-optimal. In this case, the purely economic trade-offs become less important than consideration of the net value of the mandatory investments in the fixed and recurring costs that are the price of admission to an on-call provisioning service. That is, the economic analyses in this case begin to look more actuarial than financial, forcing prospective customers to answer questions like: "Am I willing to spend X dollars a year in order to have access to a service that we can expect to need for only a few days every 3 years or so?"

For such situations, then, the driver in the users' decision whether to purchase on-call provisioning services is the cost of avoiding the effects of the expected outages and overloads. If there is a sufficiently high probability of a very serious problem, or the expected number of instances of minor problems that will be ameliorated by use of the service is great enough to produce a low cost/event ratio, the service will continue to be attractive. If not, the likely outcome of attempts to sell on-call provisioning services to that customer is the realization by the customer that his resiliency criteria are unrealistically conservative.

A.3   Quantifiers

In the development of the evaluative concepts just presented, we defined the generic measure of responsiveness of an on-call provisioning service to be the proportion of events, PS, for which:

PRT $<$ LT $+$ STT

The following sub-sections describe:

1. Formulas for estimating PRT from data on performance of the systems that establish site-to-site connections for a variety of procedures for implementing on-call provisioning; and
2. Likely ranges of LT and STT for classes of events for which on-call provisioning might be used.

A.3.1   Types of Service

As described earlier, the value of PRT for an given on-call provisioning service depends both on performance characteristics of the systems used to provide additional capacity, and on the choice of procedures for maintaining the service and responding to user requests. To accommodate and illustrate the effects that operating procedures can have on PRT, we will define formulas for estimating PRT from performance characteristics for four general classes of procedures, defined by whether maintenance and activation of on-call services are reactive or proactive. The distinctions are defined as follows:

A.3.1.1 Maintenance

1. *Reactive.* Under completely reactive maintenance procedures, necessary accesses to user sites are installed and checked out, but no further maintenance actions are taken except in response to a user notification that activated circuits are not performing properly. Upon receipt of such a request, the necessary interfaces among accesses and provider transport facilities are activated and verified, and the user is notified that the requested service is available. The user then attempts to use the circuitry turned up. If it is satisfactory, the clock stops; if not, the user notifies the provider, who only then undertakes actions to troubleshoot the circuits and correct the cause.

2. *Proactive.* Under completely proactive maintenance procedures, necessary accesses to user sites are installed and checked out, and the service provider thereafter begins to monitor those facilities, as if they were in use, to ensure that any problems are quickly detected and corrected. The intra-network links that might be used to set up connections between user sites are similarly monitored and maintained on a continuous basis.

A.3.1.2 Activation

1. *Reactive.* Upon receipt of a user request for capacity, the necessary interfaces among accesses and provider transport facilities are activated and verified, and the user is notified that the requested service is available. The user then attempts to use the links turned up. If they are all satisfactory, there is no further action; if not, the user notifies the provider, who only then undertakes actions to troubleshoot the circuits and correct the cause.

2. *Proactive.* Upon receipt of the user request for capacity, the requested links are turned up. At each step in the turn-up, the provider tests each new segment and interface while activated, verifying that they are working properly, or initiating corrective actions. As a result, each link turned over to the customer is fully functional and operating properly at the time of the turn-over.

A.3.2   *Performance Measures and Formulas*

The basic measures of performance needed to estimate PRT as a function of the types of service and performance characteristics of the facilities used to provide requested capacity are:

1. *TBF*: times between failures of segments in the site-to-site connections provided in response to user requests; and
2. *TTR*: times to restore performance across segments, when they have failed, or are malfunctioning.

Ideally, the source of these measures would be empirical distributions (or raw performance data, so that empirical distributions could be created), so that means and ranges of these times can be used in the estimates of PRT. In lieu of such empirical distributions, however, adequate estimates of the PRT can be obtained from MTBFs and MTTRs, as possibly derived from the following relationships.

We can use the measure of availability, A, of a segment, together with either its MTTR (mean time to restore) or MTBF (mean time between failures) to obtain the other, by use of the formula:

$$A = \text{MTBF}/(\text{MTBF} + \text{MTTR})$$

which shows

$$\text{MTBF} = (\text{MTTR})[A/(1 - A)]$$

and

$$\text{MTTR} = (\text{MTBF})[(1 - A)/A]$$

The MTBF for a segment can be calculated from its failure rate (F) by the formula:

$$\text{MTBF} = 1/F$$

The overall MTBF, M_o, for an end-to-end circuit comprising a series of n segments with MTBFs $M_1, M_2,...,M_n$ can be calculated from the segment values via the formula:

$$M_o - 1/[(1/M_1) + (1/M_2) + ... + (1/M_n)] = 1/[F_1 + F_2 + ... + F_n]$$

where $F_1, F_2,...,F_n$ are the corresponding failure rates.

The overall MTBF, M_o, for a pool of n interchangeable segments (e.g. IMTs) with common failure rate, F, only one of which must be fully operational and ready for use in order to be able to provide service, can be calculated from F using the formula:

$$M_o = (1/F)(1 + 1/2 + 1/3 + ... + 1/n)$$

A.3.3   Components of PRT

To develop the formulas for estimating the PRT under various operating procedures from the TBF and TTR for segments of links, it is useful to think of the PRT as comprising at least three distinct components:

- The time it takes to set up a end-to-end connection;
- The time it takes to effect correction of failures or malfunctions of segments encountered as the end-to-end connection is being set up; and
- Overheads, in the form of added delays in setting up the connection, initiating corrections when they are needed, etc.

The nature of, and nomenclature for, these three components of the PRT are described briefly below.

A.3.3.1 Route Configuration Time (RCT)

In order to set up a requested connection from user site A to user site B, the provider must:

1. Set up a long distance transport segment from the provider switch terminating user site A (SwA) to the provider switch terminating user site B (SwB);
2. Route the lines from site A to the transport segment at SwA; and
3. Route the lines from site B to the transport segment at SwB.

This part of the set up process is usually accomplished at one sitting at a network control terminal. The time to complete it is referred to here as the RCT. The RCT represents the minimum possible value of the PRT.

A.3.3.2 Segment Restoration Time (SRT)

Once an end-to-end route from user site A to user site B is configured, it will comprise three segments: user site A to provider switch SwA; the intra-network transport segment from SwA to SwB; and provider switch SwB to user site B. In the event that the facilities assigned to implement any of these three segments have failed or are malfunctioning, it will be necessary to effect a correction. The times required for such corrections, measured from the first action that initiates a correction to activation of a properly functioning segment, are referred to here as SRTs.

A.3.3.3 Process Overhead (PO)

The RCT plus the expected contributions of delays due to needs for segment restorations reflected in SRTs represents, by definition, the least possible PRTs for the conditions encountered. In actual operations of the on-call provisioning service, there will be numerous other sources of delays that may increase PRTs from this theoretical minimum. Possibilities include, for example: delays in recognizing that a segment is not working properly; time spent checking out routes as, or after, they are configured; lapses in communications or personnel attention that create a delay in initiating the response to a user request; and elongation of RCTs due to slow-downs in the entry of information, times spent waiting for access to a terminal or the network control systems, or lack of proficiency in the entry process. Such times need not be considered separately, except to the extent that they will differ substantially with the type of operating procedures. Those that are considered must be named and defined as necessary; all the rest can be lumped into a single time added to a theoretical minimum to estimate the PRT. This single adjustment factor is referred to here as the process overhead (PO).

A.3.4   Calculation of PRTs for Different Service Types

PRTs for each of the four types of service can be expressed in terms of the components of PRT just defined and the probabilities that segments picked in the first attempt at route configuration will be malfunctioning. The appropriate differences in the calculations are as follows.

A.3.4.1 Segment Failure Probabilities

In configuring a route, there is always a chance that segments picked in the first attempt at route configuration will be malfunctioning. The probabilities of this happening are referred to here as segment failure probabilities and denoted here $P_f[s]$, where is the identifier for a segment. The segment failure probability for a given segment depends both on the performance characteristics of the facilities used in the segment and on the type of maintenance procedure employed, as follows:

Reactive maintenance. For segments that are maintained reactively, the operational condition of the segment is not checked until it is needed to configure a route in response to a user request. As a consequence, when such a segment fails or malfunctions, it remains in that state until the next time it is needed in a route configuration. The probability that it will be found in a failure state thus depends directly on:

1. The failure rate for the segment (F); and
2. The time lapsed since the last time the segment was known to be functioning properly (T_c).

Under the assumption that the failure rate is constant over the period involved, the failure probability for the segment is given by:

$$P_f = 1 - \exp[-(F)(T_c)]$$

where $\exp[x]$ denotes the exponential function of x.

Proactive maintenance. For segments that are maintained proactively, the state of the segment is continuously monitored and the segment is maintained exactly as if it were carrying traffic. The performance characteristics of the segment will, therefore, closely approximate the operational performance characteristic, and the failure probability will be given by

$$P_f = 1 - A$$

where A is the availability of the segment.

A.3.4.2 Calculation Formulas

The way that a particular set of failure probabilities and components of PRT should be combined to estimate PRT further varies with the activation procedure employed as follows:

Reactive activation. When site-to-site connections are activated reactively, one of two outcomes prevails. Either all the segments in the connection were functioning properly when the link was turned up, and the PRT approximates the RCT, or one or more of the segments turned up is not functioning properly, and corrective actions are required. The probability that the latter case will prevail, PF, can be calculated directly from the three segment failure probabilities from the formula:

$$PF = 1 - (1 - P_f[a])(1 - P_f[b])(1 - P_f[t])$$

where a, b, and t denote, respectively, the site A/SwA segment, the site B/SwB segment, and the intra-network transport.

In the event that the need for correction does prevail, the process overhead will include, in addition to other process delays that may be experienced with proactive activation, a delay in discovery of the condition that results from the fact that the route is configured, but not tested until the user tries to use it. This creates a latency in the problem that approximates the difference in LT (the lead time provided by the user) and RCT. If we denote the problem latency by LAT and incorporate it into the test conditions for adequacy of the provider response, we get the relationships:

$$PRT = RCT + PO + SCT + LAT$$

and

$$LAT = LT - RCT$$

Substitution into the condition for adequacy of the provider response that
PRT < LT + STT, and simplification produces for this case the derived
condition that:

PO + SCT < STT

Since this is an unlikely condition, requiring that all the process overheads
and the maximum service restoration time to be less than the shortfall toler-
ance time, the best conservative estimate of satisfaction of user requirements
under this activation procedure is to assume simply the only time that user
requirements will be satisfied is when all segments are fully functional when
they are turned up, and set the measure of responsiveness to:

1 − PF

Proactive activation. When a route is configured proactively, and the necessary on-line test capabilities are available, the procedure is to test the segments
before, or as, they are configured to produce the end-to-end link. When no
corrections are necessary, this procedure increases the PRT by adding the time
needed to test each segment. When corrections, rather than selection of alternate segments are necessary, this procedure adds an additional delay in the
form of the time needed to restore any failed or malfunctioning segments. As
in the case of reactive activation just described, the probability that one or
more corrections will be needed is PF, defined by the same expression invol-
ving the individual P_fs. Since all problems will have been detected within a
short time of each other, the additional time added to the RCT in completing
the activation will be the longest time to restore among the TTRs for the failed
segments.

The PRT thus becomes:

RCT + PO + TO with probability 1 − PF

and

RCT + PO + TO + max{SRT(a), SRT(b), SRT(t)} with probability PF

where TO denotes the total time expended in testing the segments, SRT(.) are
the segment restoration times for each of the three segments as defined earlier,
and max{X} denotes the maximum value of the set of numbers X.

These results immediately imply that for services in which the activation of
capacity in response to user requests is accomplished proactively, there are but
three possible measures of responsiveness:

1. RCT + PO + TO + max{SRT(a), SRT(b), SRT(t)} < LT + STT, in
   which case the service is almost completely responsive;

2. RCT + PO + TO < LT + STT, in which case the measure of responsiveness is 1 − PF; or
3. RCT + PO + TO > LT + STT, in which case the service is completely unresponsive, regardless of the value of PF.

Appendix B

A Short Lesson in How *Not* to Measure Quality of Service

The following is a dramatization of a conversation that I have had with marketers over the last two decades more times than I would like to count. The exchange invariably begins with someone calling me up to see if I have, or can produce figures on the availability of our services that some prospective customer wants to review before making a purchase decision. My standard retort to this question is one of my own:

> Why does your customer want to know? "Availability" is worse than worthless as a metric for telecommunications quality of service!

From this low-keyed challenge, the discussions runs something like this, with me, as the analyst, A, responding to the marketer, M.

M: How can you say that? Whenever I get a chance to talk to a prospective customer, one of the first questions that comes up is availability. And, more and more often it seems we are asked to provide some sort of penalties for not living up to whatever availability figures we quote. I have got to be concerned about availability.

A: Perhaps. But you must realize that every time you quote an availability figure to a customer, you are responding to a problem that has been expressed as a solution. Unless the customer is very sophisticated, that request for information on "availability" is that customer's way of asking you for reassurance with respect to the real concern.

M: And that "real" concern is...

A: ...how often the service will be inaccessible long enough to materially affect the company's operations.

M: Isn't that what availability measures?

A: Yes and no. The problem is that availability as a single number reflects two entirely different phenomena. Availability is a ratio of time up to total time, calculated over some time interval. What determines that ratio, however, are two factors – the number of outages that occurred, and the time it took to recover from those outages. Agreed?

M: Well, yes, but so what?

A: The problem, then, is that because two independent effects are represented in the ratio, achievement of a particular level of availability may or may not assure that the conditions feared by the customer will eventuate with unacceptable frequency. It is all in how the ratio is determined.

M: Show me what you mean. I still do not get it.

A: OK. Let's consider this. Suppose a customer expresses a concern that the availability of a particular service should be 99.9% or better. That is a relatively low availability number for telecommunications service like a business telephone line or a dedicated termination for handling 800- numbers, but let's work with it for purposes of illustration. Now, when you "guarantee" the customer that the availability will be 99.9% or better, what are you promising?

M: That if we calculate the ratio (time the service was up)/(total time operated) the number will be 0.999 or better.

A: Well, more accurately, you must be promising that this will happen when the ratio is calculated over a long enough time interval to yield a sufficiently accurate ratio. But I will get to that part in a moment. What I want to point out here is that when you satisfy that condition, the customer may have experienced:

- An 11 s outage once every 3 h
- A 1.4 min outage once a day
- A 10 min outage once a week
- A 22 min outage twice a month
- A 45 min outage once a month
- An 8.5 h outage once a year
- A 17 h outage once every 2 years
- A 24 h outage once every 2.7 years
- A 42 h outage once every 5 years

This means that whether the 99.9% availability achieves the customer's ends depends on what constitutes a perceived threat to the customer's operations. If the customer's nightmare is an outage that lasts long enough to put the office out of business for a day, the thought of an 8.5 h outage once a year is not likely to be reassuring. However, if the business use of the service is a regular office environment, that 10 min outage once a week is no big deal, because the problem will be transparent to most of the users – unless, of course, there are a large number of users who are using the service for Internet log-ons that time out after 5 min of inactivity. In that case, the 10 min outage once a week might have the users ready to lynch the comm manager, while the 1.4 min outage once a day, or the 17 h outage once every 2 years would be largely transparent, and tolerable, to the users...

M: ...unless every outage results in automatic disconnection of all calls in progress, in which case the 11 s outage once every 3 h is a nightmare, even if that system almost never experiences a long outage.

A: Precisely. Or maybe I should say "imprecisely". The point here is that an availability figure by itself almost never addresses the customer's concern as to whether the outages users can expect to experience will be tolerable. What is tolerable depends both on the expected frequency of outages and their duration, and that information is lost in the calculation of availability.

M: So I am doing my customer a disservice whenever I simply assert that our service will achieve the specified availability, without understanding the underlying concern and explaining what that availability figure means with respect to that concern.

A: Yes, and you are also doing yourself a large disservice. Once you have given that answer without explanation, you are leaving yourself open to later problems that can seriously undermine, if not destroy, your credibility with that customer. Do you remember how I said that when the availability ratio is calculated, it is very important to be sure that the total operating time sampled is long enough to assure that the ratio is a sufficiently accurate estimate of the underlying availability?

M: Yes. And I can readily see that we had better average things out over a long time period, or we are likely to come up with an artificially low estimate of availability.

A: For example, suppose we estimate availability based on 3 months of operations. If there were a quarter that had an outage, preceded by two or three quarters during which there were no outages at all, the estimate of availability for the quarter with the outage would be much lower than we would expect on the basis of three quarters' experience.

M: And that would mean that I am going to be called to task for the fact that the expected availability was not met during that quarter, even though the average over the longer period showed that the availability was just what I told the customer it would be.

A: Yes. Just imagine having to dance around that one when you have promised the customer that the availability averaged over any 3 month period would meet the specification. But, it can and will get even worse. As the technology gets better and better, we are going to see lower and lower failure rates, getting availabilities that result from very infrequent outages that may nonetheless take some time to correct. So imagine this. You have promised the customer an availability of 99.9%, which is realized on average as one outage every 2 years lasting on average 17 h. After the service has been up and running for, say, 6 months, you have the first outage, which unfortunately lasts 36 h. When your customer calls you in and wants to know if this is the kind of service to expect, what are you going to say?

If you try to explain that this is only one of the expected events, and other subsequent outages run the average down to the expected 17 h, and the customer asks how long it will take to verify that the availability is as advertised, your answer has to be something like, "Oh, we can show that for your service in another 2–10 years. But to be certain that we are not below the advertised availability, it will really take something like 30–60 years to get a sample big enough to prove it". Imagine your customer's reaction to that one!

M: OK. So it is difficult to be clear and safe about all of this. I have still got to respond to customers who want to know about availability of our various services. What can I do?

A: The best defense is to educate, educate, educate. When you interact with a customer who is asking about availability, do not simply feed back a number or guarantee that you think will be satisfactory. Probe the customer's needs to determine what is behind the concern with availability and what outage characteristics are particularly painful. What customer activities are affected by outages? How many outages of any kind are likely to test users' patience? How long can an outage endure before it materially affects user perception of quality of service? In short, work with the customer to identify the underlying concerns, bring them to the surface, and show what levels of availability will assure that the painful circumstances will be avoided, or, at least, be tolerable. If necessary, show the customer how the real concerns with availability of particular services can be addressed by use of redundancy, diversity, etc. to achieve the desired operational availability, even though a single threaded service cannot.

M: That is fine if I can open the dialog. But what about the responses required to open the door? With a lot of customers, it is impossible to enter into discussions until you have responded to the basic specifications.

A: Well, the easiest answer to that is that there is no restriction on how you should respond to the basic specifications. What I would do, then, would be to respond to a customer request for availability figures by displaying the OCC for mean times between outages, and pointing out that the data summarized there shows that the availability is such-and-such. Given the opening of using the OCC as the reference for the availability ratio, you might also just slip in a few indications of what else the curve shows, such as how often we expect outages lasting 4 h or more as compared to outages lasting 2 h or more. This changes your original written response to the question about availability from a one line answer to a one page answer with a figure, but I would be very surprised if the response did not elicit a lot of interest in what is being displayed.

M: And if it doesn't?

A: You are certainly no worse off than when you started...

Appendix C

Problems with Interpretation of Answer-Seizure Ratios (ASRs)

The answer/seizure ratio (ASR) is a measure widely used as an indicator of performance in the international community. For given sets of origins (O) and destinations (D) it is calculated as the ratio:

$$\text{ASR}(O, D) = N_a(O, D)/N_s(O, D)$$

where $N_s(O,D)$ is the number of call attempts to destinations D from origins O, and $N_a(O,D)$ is the number of those call attempts resulting in an answer from a distant station.

Use of this ratio as an indicator of performance is implicitly predicated on three assumptions:

1. The calculated value of ASR is an accurate, meaningful estimate of the probability that a call will be answered (P_a).
2. $P_a = (P_c)(P_{a/c})$, where P_c is the probability that a call attempt will be normally completed (i.e. result in a distant station ring or busy signal), and $P_{a/c}$ is the probability that a normally completed call attempt will be answered.
3. $P_{a/c}$ is a stable characteristic of normally completed call attempts from a set of origins to a set of destinations.

Under these assumptions, variations in P_a are dominated by P_c, which implies that the ASR varies directly with P_c, and supports inferences to the effect that changes or differences in ASRs reflect changes or differences in connection reliabilities.

The problem in using ASRs to monitor service or compare performance of different services, however, is that there are many ways in which one of these assumptions can be violated to produce fallacious inferences from analysis of ASRs. In particular, there are four possible pitfalls in such analyses, characterized generally as stemming from:

1. *Incommensurate quantification of ASRs*: differences in the data that are used to calculate ASRs that may result in incommensurate values.
2. *Inadequate sample sizes*: failure to base calculation of ASRs on samples that are large enough to produce stable estimates of the probability that a call attempt will result in an answer.
3. *Sample inhomogeneity*: failure to assure that samples from which ASRs are calculated are comparable with respect to all the major factors that can affect the probability that a call will be answered.
4. *Misattribution of causes*: failure to account for all possibilities for explanation of significant differences among ASRs.

The intent here is not to argue that conclusions derived from analyses of ASRs are necessarily wrong, suspect, or useless, but to demonstrate the need to be very careful in interpreting comparisons of ASRs when those results may affect decisions for which there is a high cost of being wrong. When inferences from analyses of ASRs are proffered as the basis for some expensive action, such as re-allocation of limited resources, development of new routes, etc. the possibilities for erroneous inferences described below mandate extreme care in interpreting the data and validating the conclusions, to minimize the chances that a multi-million dollar decision will be based on bad information. Other uses of ASRs, however, may not require such precautions. For example, when ASRs are used as a means of identifying possible changes in performance of a service, the decision supported is one of where to look for possible problems, and the indications from analysis of ASRs provide information that might otherwise be unavailable or submerged in masses of maintenance data. The only penalty for a wrong decision in this case is the effort expended in pursuing what turns out to have been a false alarm, and this penalty will be adequately off-set by the benefits of pursuing indications from analyses of ASRs, as long as a sufficient proportion of those indications do surface latent problems.

C.1   Incommensurate Quantification

In order to meaningfully compare two ASRs, the ratios calculated must be mutually consistent in the sense that their numerators and denominators reflect counts of identical phenomena determined by the same criteria. Because of the widespread use of ASRs without standards for the data bases from which they are calculated, however, this requirement may not be satisfied, so that two agencies can produce significantly different estimates of ASRs for the identical traffic. For example, there are in general two different kinds of data bases that might be used to calculate ASRs:

1. Call records comprising the data used for billing purposes; and
2. Call records comprising the data on call attempts logged on particular trunk groups through which all traffic from a set of origins to a set of destinations is routed.

The ASRs calculated for the same set of origins and destinations from these two different data sources will necessarily differ by a factor determined by the proportion of call attempts that did not result in a seizure of the trunk groups monitored.

More generally, there are numerous ways in which values of ASRs may be affected by variations in the data bases used to calculate them, producing differences in values that reflect inconsistencies in the data rather than a difference in performance. As suggested by the example above, the key to detecting and understanding such differences is to precisely answer the question: "What is counted as a seizure?" The answer may reveal subtle differences, even when the ASRs are calculated from the same data base. For example, in the case of a primary service and an alternate which handles overflow from the primary service, the slightly longer time required to route a call to the overflow will result in proportionally fewer user-abandoned calls that result in seizures of the overflow trunks. Thus, what is counted as a seizure for either service may not be the same, even though the mechanism for logging a seizure is identical for both services, and it becomes necessary to determine whether that difference has any substantial impact on the relative values of ASRs before using those results to infer differences in performance.

C.2   Inadequate Sample Sizes

As indicated above, one of the principal assumptions on which use of ASRs to analyze performance is based is that the ASR represents an accurate estimate of P_a, the probability that a call attempt will be answered for the set of origins and destinations examined. If it is assumed that P_a is a stable probability (i.e. the same for any random sample of call attempts), then the size of the sample required to estimate that probability grows rapidly with the accuracy required. For example, the following table shows the approximate sample sizes required to establish with 97.5% confidence that the ASR approximates a stable P_a with the indicated accuracy.

Accuracy ($\pm$)	Required sample size
0.10	105–125
0.075	187–223
0.05	421–502
0.04	659–784
0.03	1171–1394
0.02	2634–3136
0.01	10536–12544
0.005	42148–50176

This table shows, for example, that if we calculate two ASRs, one at 60% based on a sample of 400–500 call attempts, the sample size for calculation of a second ASR at 66% would have to exceed 12 544 call attempts before we could be 95% confident that the observed differences were not due to chance fluctuations in the data.

Moreover, the sample sizes are predicated on the assumption that P_a is a stable probability (i.e. the same for any random sample of call attempts). In fact, it is much more reasonable to assume that P_a for any given sets of origins and destinations is not stable, but varies, as a minimum with time of day. If this is the case, the required sample size increases approximately as the square root of the number of different states for which the value of P_a is stable. Thus, for example, if we were to find that there were different values of P_a during six different 4-h time periods during a day, the required sample sizes shown above would be increased by about 2.5 times, and we could not conclude that the observed difference in the example was significant unless the first ASR was based on a sample of 1000–1250 call attempts and the sample for the second ASR was based on more than 31 000.

The message here, then, is that comparisons of commensurate ASRs based on any but very large sample sizes may exhibit substantial differences that are, in fact, not indicative of material differences in performance, even when they pass tests of significance based on the assumption of stable values of P_a. The only way to avoid such errors is to verify the stability of P_as, or to base comparisons of ASRs on distributions of the ratios for small numbers of call attempts rather than a large sample average.

C.3   Sample Inhomogeneity

No matter how accurately ASRs are calculated they only estimate the value of P_a. Consequently, their utility as indicators of performance is limited unless $P_{a/c}$, the probability of an answer given a normal completion (i.e. ringing or a station busy signal) is stable and relatively constant for the samples on which the ASRs being compared are based. Unless this condition is satisfied, comparisons of ASRs will reflect comparisons of P_as without supporting inferences as to the relative values of the corresponding P_cs. In the case of international telephone services for which ASRs are used, there are at least five major factors that will affect the unconditional probability that a call will be answered:

1. Country called, since the performance of domestic networks that complete international calls varies widely from country to country;
2. Geographic distribution of destinations called, since performance characteristics of domestic telephone services vary with the network structure, even in countries with very good telecommunications services;
3. Time of day, since the probability of getting an answer, given a normal completion depends on the likelihood that a person or machine is present to answer the telephone, and the probability of blocking in all domestic networks varies with diurnal variations in traffic volumes;
4. Type of location called, as distinguished, for example, by whether the station called is in a private residence, a business office, or a telephone center comprising many stations, at least some of which are guarded continuously during certain hours (e.g. a freephone answering or operator service centers); and
5. Type of call, as distinguished by whether the station is answered by humans, modems, or telephone answering machines.

Of theses, factors (3), (4), and (5) also affect $P_{a/c}$, creating at least 18 combinations of time of day, type of location, and type of call for which the stable values of $P_{a/c}$ (if they exist) can be expected to be different. In practical terms, this implies that inferences as to relative performance derived from comparisons of ASRs can easily be erroneous, unless it can be verified or assured that the samples of call attempts from which they were calculated contain approximately the same proportions of calls in each category. This requirement is presumed to have been met when comparisons of ASRs are based on very large samples of call attempts recorded continuously over a long time interval. However, it is not readily apparent that even a 28-day sample of all call attempts from a particular set of origins to a particular set of destinations will automatically satisfy the requirement for homogeneity of samples, and verification or other forms of assurance may be necessary to produce a convincing argument that observed differences in ASRs reflect substantive differences in P_c.

C.4 Misattribution Of Causes

Notwithstanding the other impediments to meaningful comparisons of ASRs, there is always a danger that the proximate cause of a significant difference between two ASRs may not have anything to do with circuit capacities, quality of signaling interfaces, reliability of equipment, or other characteristics that commonly affect P_c, but is attributable to something not suggested by the ASR model.

The following examples illustrate this point. The mechanisms described are mathematically sound, even though it is unlikely that the resultant differences would be detectable in comparisons of ASRs except in extreme cases.

1. Given two services with identical P_cs, the one offering inferior voice quality may have the higher ASR. The reason is that, all other factors being equal, the sample of call attempts for the inferior service will have a higher proportion of calls for which $P_{a/c} = 1$, because the calling party, having reached the called party opts to drop the call and replace it in the hope of getting a clearer connection.
2. Given two services with identical P_cs, and all other factors being equal, the one with the higher premature disconnection rate will similarly have a higher ASR, because of a higher proportion of calls re-placed to stations where the persons disconnected are waiting for a call back.
3. When ASRs are based on billing data, so that all call origins are reflected in the denominator, greater post-dial delay and consequent higher user abandonment rates will reduce the ASR for a service without any change in other performance characteristics that affect P_c.

Abbreviations

Δ**[SO,SI|ty]**   the average difference between duration of service outages and duration of perceived service interruptions as a function of the type of service outage denoted by ty

AC[t]   accessibility distribution function: the probability distribution function for operational service interruptions defined, for example, so that when t is time measured in hours, AC[t] represents the probability that an operational service interruption will last t hours or longer

ADR   abnormal disconnect rate: the proportion of observed connections for which a transaction was initiated, but the circuit was disconnected before the transaction was complete

ANS   station answer: the answer received when a call attempt is completed to the station dialed

ASR   answer-seizure ratio: the proportion of call attempts resulting in line seizure at a certain point that are answered by a person or device at the station called

ATM   Asynchronous Transfer Mode: a packet-switching protocol used for establishing connections via very high data rate optical communications links

AV[t]   availability distribution function: the probability distribution function for service outages in intermittently used services defined, for example, so that when t is measured in hours, AV[t] represents the probability that a service outage will last t hours or longer

APL   A Programming Language: an early, but very versatile interpreted computer programming language distinguished by its use of strange looking characters to tokenize calls to macros implementing frequently used mathematical functions and routines

BER bit error rate: a measure of the fidelity of transmissions over a data link expressed as the proportion of all bits transmitted that are received in error

CA (no fixed meaning): an convenient mnemonic for a parameter variously used in different contexts as an abbreviation for: the average duration of connection attempts effected; number of connection attempts answered; connection availability; and number of call attempts

CCR call completion rate: the proportion of call attempts that result in a verifiable connection to the distant station called

CCS7 Common Channel Signaling System No. 7: the system for out-of-band signaling most commonly used today to effect routing of origin/destination connections through public switched networks

CDMA code division multiple access: a multiplexing scheme whereby transmission capacity is divided both by frequency and time, and virtual channels are created by multiplexing codes that allocate different frequencies to transmission of data during different time slots

CELP code-excited linear predictive coding: an encoding scheme used in CODECs whereby segments of electrical analog signals are processed to determine the best fit to a library of segments of waveforms and the digitally encoded symbol for each segment is transmitted to the distant end, where each segment is reconstructed according to the description transmitted

CODEC coder/decoder: a device for digitizing analog electrical signals for transmission and re-constructing the analog signals from the digital data received at the distant end

DAL direct access line: a circuit that runs directly from a long distance service switch to customer premises, without intermediate switching through a local service switch

DER data error rate: a measure of the fidelity of transmissions over a data link expressed as the differences between the injected data at the origin comprising what is to be delivered and the image of that data extracted at the destination

DEOT direct end-office termination: a circuit that runs directly from a long distance switch to a local end office, without intermediate switching through other local service switches

DBMS   data base management system: a computer software package that implements routines for definition, construction, manipulation, and queries of databases

DPR   dropped packet rate: the proportion of packets transmitted via a packet-switched service that are not received at the destination on the first transmission attempt

DRR   disconnect report rate: the proportion of calls completed by a group of users for which the service provider receives user complaints that the connection was taken down before the parties to the conversation were ready to hang up

DS0   digital signal level 0: a 64 kbps digital data channel; capable of carrying a voice channel digitized with 8-bit PCM

DS1   digital signal level 1: a 1.544 Mbps digital data channel; employs time division multiplexing to carry up to 24 DS0s

DSx/DSy multiplexer   a device for multiplexing digital signal level x channels by packing them into digital signal level y channels, $y > x$; a DS0/DS1 multiplexer, for example, packs up to 24 DS0s into one 1.544 Mbps DS1 channel

DTMF   dual tone multiple frequency: the audio signaling system used by push-button telephones wherein each number on the key pad is associated with a waveform comprising the sum of two different frequencies transmitted at the same level

E1   the European version of a T1 channel; utilizes time-division multiplexing of a 2.048 Mbps data signal instead of the 1.544 Mbps used for T1

EDR   effective data rate: the equivalent rate of transmission of data achieved over a connection after throughput efficiency, handling overhead, and encoding overhead have been accounted for; for a data transmission rate, d, the EDR is the ratio: $[(d)(\text{TE})]/[(1+\text{HO})(1+\text{EO})]$

EO   encoding overhead: the amount of data that must be added to an information transmission unit for identification, framing, sequencing, transmission control, or forward error correction, expressed as the ratio of the amount of that data to the unformatted size of the information transmission unit, so that the size of unit is inflated by the factor (1+EO)

EoS   economy of service: a notion that is as intuitively self-evident, but formally elusive, as QoS

FAX facsimile: transmission of images over telephone lines accomplished by scanning material to create digital images and transmitting the results of digitization via ordinary telephone lines by use of modems that encode digital data into analog electrical signals; the encoding alternatives and standards are set by the ITU V-series of recommendations for transmission of data over the public switched network

FDMA frequency division multiple access: a multiplexing scheme whereby transmission bandwidth is divided into channels defined by different frequencies which are allocated to carry the information transmitted over an assigned connection

FORTRAN FORmula TRANslator: one of the earliest compiled computer programming languages, designed by IBM for scientific applications

H/D high-and-dry: an outcome of a call attempt in which nothing is heard after what is perceived to be such an inordinately long time that the user abandons the attempt

HO handling overhead: the amount of information that must be added to an information exchange unit to support proper routing, handling, and delivery of units through a continuously-used service in which the originator cannot actively oversee transmission; expressed as the ratio of the amount of handling information to the amount of injected information, so that the volume of injected information is inflated by the factor (1+HO)

HSR hand-shake success rate: the proportion of data modem/FAX call attempts that result in synchronization and initiation of data transmission after being answered

IP Internet Protocol: the routing and transmission protocols that implement the collection of interconnected packet-switched networks called the Internet

ISDN Integrated Services Digital Network: a telecommunications service that carries both voice and data over the same facilities

ITU International Telecommunication Union: a body of the United Nations devoted to achieving cooperation among, and interoperability between, the telecommunications service providers of different nations

kbps thousand bits per second

Mbps million bits per second

MERS most economical routing system: a software package installed on

PBXs through which outgoing calls are analyzed and routed to the service alternative among several accessible via the PBX that offers the lowest price

MOS   mean opinion score: a measure of quality of voice services obtained by assigning numerical values to qualitative descriptions of quality, e.g. "excellent", "good", "fair", and finding the average of the numerical scores so assigned

MTBF   mean time between failures: average time between consecutive failures of a system or piece of equipment

MTTR   mean time to repair (restore): average time it takes to recover from a failure of a system or piece of equipment; restoration may entail reversion to a back-up rather than repair of the original source of the failure

MTBOSI   mean time between operational service interruptions: average time between consecutive service outages that are detected by users as operational service interruptions

NCR   normal completion rate: the proportion of call attempts that result in a ring-back, a station busy signal, or an immediate answer without ring, thereby being manifested to users as normal outcomes

NNX   the second group of three digits in the North American telephone numbering plan; each of NNXs used in any area is assigned for use by only one telephone exchange end office

NPA   the first three digits, comprising the area code, in the North American telephone numbering plan

OCC   operating characteristic curve: a graph displaying a probability distribution function in which values on the x-axis are times, representing durations of events of some kind, values on the y-axis are mean times between occurrence, and the value of y displayed for any value of x is the expected time between consecutive occurrences of events lasting x time units or longer.

OEC   operational effective capacity: a measure of accessibility of continuously-used services, defined generically to be the expected ratio of user-injected information exchanged over some time period to the maximum rated capacity of the service for exchange of data or characters

OEG   Operations Evaluation Group: a sub-division of the Center for Naval Analyses whose principal activity is direct analytical support of tactical and operational decision-making for US Navy commands

OSI   operational service interruption: an interruption of a telecommunications service that is both unexpected and detected by the users of the service

PBX   private branch exchange: a small switch deployed to handle routing of calls into and out of a customer premise or campus

PCM   Pulse Code Modulation: an encoding scheme used in a CODEC whereby the amplitude of an analog electrical signal is sampled at fixed intervals; the values of the magnitude of the amplitudes are digitized for transmission, and the digital values received are used to regenerate the analog signal at the distant end

PDD   post-dial delay: the time lapsed between the dialing of the last digit in a telephone number and detection of the first network response; usually quantified by timing only the call attempts that resulted in ring-back or answer by the station called

PP[x]   probability that a service outage of duration x or longer will be perceived as an operational service interruption

PR[i]   probability of experiencing a run of exactly i consecutive failures of call attempts to the same destination; PR[0] denotes the probability of success on the first call attempt

PRPDD   primary route post-dial delay: the post-dial delays represented by the cluster of values about the smallest significant mode in the frequency distribution of PDDs for a set of particular origin/destination connections

P[UDI]   a measure of the quality of a voice service reflecting the proportion of calls tested in a Service Attribute Test (SAT) for which the effects of the impairments were reported to have rendered the call "unusable", "difficult", or "irritating"

QoS   quality of service: an intuitively self-evident, but formally elusive, concept that is the subject of this book

RCT   ringer connect time: the time it takes to connect a ring signal generator to the origin/destination circuit after the line to the destination station has been seized

RDR   reorder (network busy) signal: an audible signal, usually transmitted at 120 ips, indicating that the requested connection cannot be made because there is no available facility for effecting one of the node-to-node links needed to complete the connection, or the routing information is incomplete

RNA   ring, no answer: a condition manifested when a connection is completed to a distant station, but no person or device answers

RNG   ring back signal: the audible signal indicating that the connection has been set up to the destination station; usually, but not necessarily, generated with a cadence of 4 s of silence followed by 2 s of signaling comprising one or two pulses of a tone

RRR   retransmission request rate: the proportion of messages delivered via a message relay service for which the recipient requests a retransmission from the originator because the message has been received with undecipherable errors or missing elements

RSL   ring signal latency: the expected amount of time after a ringer has been attached before a ring signal will be audible

RVA   recorded voice announcement: a voice recording played in response to a ring signal explaining the condition that is preventing completion of the requested connection to the desired party

SAT   Service Attribute Test: a subjective user test conducted under a particular protocol which calls for test subjects to place repeated calls to cooperating destinations, hold short conversations, and report in addition to their opinions of the quality of each call the incidence and severity of particular impairments, together with a quasi-objective description of the overall effects of those impairments

SBS   Satellite Business Systems: a now defunct corporation that developed and sold commercial telecommunications services utilizing geostationary satellites for signal transport; sold public switched telephone services under the brand name "Skyline"

SBY   slow (station) busy signal: an audible signal, usually pulsed at 60 ips (impulses per second) indicating that the distant station called is already in use or "busied out" by the local service provider

SIT   special information tone: an audible signal comprising a three-tone warble indicating some problem with the number dialed making it impossible to route; frequently followed by a recorded voice announcement indicating that the call could not be completed as dialed and citing a reason

SLA   service level agreement: a contract between a telecommunications service provider and a customer in which the provider agrees to reductions of cost or other monetary compensation to the customer whenever it is demon-

strated that the level of quality of service with respect to particular characteristics fail to meet specified criteria

SONET Synchronous Optical NETwork: a protocol for establishing telecommunications links via modulation of lightwaves transmitted through fiber
optic cables

SQL Structured Query Language: one of standard query languages specified
for use in data base management systems (DBMSs)

SSR service seizure rate: the frequency of line seizures/log-ons onto a telecommunications service, expressed as the average number of occurrences per
unit time

SUT subjective user test: a test of voice quality conducted by asking users
their opinions of quality of calls placed via the service(s) being tested

TDMA time division multiple access: a multiplexing technique whereby
transmission capacity is divided into time slots and the signals from multiple
sources are transmitted by allocating sets of time slots for use by each source

TE throughput efficiency: the ratio of the number of information transfer
units actually exchanged via a continuously-used service over a time period
when the service was fully available to the maximum number that could have
been exchanged were the connections via the service completely error-free

T1 another name for a DS1 facility; transports up to 24 64 kbps digitized
voice channels by using time-division multiplexing of a 1.544 Mbps data
stream

UPR use-to-potential ratio: a measure of the utilization of a group of circuits
obtained by dividing the actual usage by the potential maximum usage consonant with the assumption of non-blocking of an assumed pattern of variations
in traffic loads

VRU voice response unit: a device for automatically answering incoming
telephone calls with a recorded voice announcement (RVA)

V.x ITU V-series no. x: a general reference to ITU recommendations for
transmission of data over the public switched network

Index

Notes: page(s) on which an item is defined are denoted by inclusion of 'd.' before the page number. Other abbreviations used are: il., denoting a reference to something that illustrates the item; tab., for designating a table as the location of the item; fig., for designating a figure as the location of the item.